DINOSAUR IMPRESSIONS

POSTCARDS FROM A PALEONTOLOGIST

PHILIPPE TAQUET

Translated by
KEVIN PADIAN

CAMBRIDGE
UNIVERSITY PRESS

PUBLISHED BY THE PRESS SYNDICATE OF THE UNIVERSITY OF CAMBRIDGE
The Pitt Building, Trumpington Street, Cambridge CB2 1RP, United Kingdom

CAMBRIDGE UNIVERSITY PRESS
The Edinburgh Building, Cambridge CB2 2RU, UK   http://www.cup.cam.ac.uk
40 West 20th Street, New York, NY 10011-4211, USA http://www.cup.org
10 Stamford Road, Oakleigh, Melbourne 3166, Australia

Originally published in French as *L'Empreinte des dinosaures*
by Éditions Odile Jacob 1994
and © Éditions Odile Jacob

First published in English by Cambridge University Press 1998
with the help of the French Ministry of Culture
English translation © Cambridge University Press 1998

Printed in the United States of America

Typeset in Meridien 10/13.5 pt, in QuarkXpress™ [MG]

A catalog record for this book is available from the British Library

Library of Congress Cataloging-in-Publication Data
Taquet, Philippe.
    [L'Empreinte des dinosaures. English]
    Dinosaur impressions : postcards from a paleontologist / Philippe Taquet ;
    Kevin Padian, translator.
      p. cm.
    Includes bibliographical references and index.
    ISBN 0-521-58372-1 (hb)
      1. Taquet, Philippe. 2. Dinosaurs. 3. Paleontology. I. Title.
QE22.T36T3713   1998
567.9–dc21   98–11560
                                                          CIP

ISBN 0 521 58372 1 hardback

CONTENTS

Would there not also be some glory for Man in knowing how to broach the limits of time, and through his observations to rediscover the history of this world and the succession of events that preceded the birth of humankind?

> Georges Cuvier, *Discourse on the Revolutions of the Globe*, 1825

It seemed that I had always lived like this, and I wanted it to last forever. I wanted the unknown world to be without limits and, each day for countless years, for the dragons of my tent to rear up in the air of a new land. To travel like this is to live two lifetimes; to stop, to remain, is to live half-dead. In days gone by the voyages were long. Marco Polo's lasted twenty-seven years. Those were the days!

> Jacques Bacot, *Tibet in Revolt*, 1912

TRANSLATOR'S PREFACE

LATE IN 1994 I received one of those pleasant surprises that seem to occur with uncanny regularity in the field of paleontology. My old friend and colleague, Philippe Taquet, had sent me a copy of his new book, *L'Empreinte des Dinosaures: Carnets de piste d'un chercheur d'os* (literally, *The "Imprint" of Dinosaurs: Field Notebooks of a Bone Hunter*). I read it through and was delighted by it. My first thought was that it should gain wider exposure than it was likely to get, inasmuch as Philippe's French publisher, Odile Jacob, didn't usually advertise heavily in our field. I called Nick Fraser, who edited the *Journal of Vertebrate Paleontology,* and asked if I might review it for him. He quickly agreed, and the review was soon published.

Several months later, though, I began to think that Philippe's book might not receive its due attention outside France, and that would be a shame. Unfortunately, even in our field, where people from all over the globe regularly travel and work together, we don't read in each other's languages as much as we used to, or should. Sometimes the newest or the most flashy recent discoveries have a way of dimming the memory of the work of explorers who laid the groundwork or discovered many of the same things long ago – and, as you will quickly see, Philippe has been many places and done a lot of things. Besides, he has some wonderful stories to tell. Moreover, we need his Gallic perspective, and others like his, so that our science and others don't become dominated by the historical accident of anglophonic hegemony in the conferences and publications of the scientific community.

With Philippe's acquiescence we urged Éditions Odile Jacob to have the book translated, and in due time this was arranged. When Philippe settled on Cambridge University Press he very kindly asked me if I would finish what I had started. It has been an enormous pleasure to help to bring to a wider audience the exploits and insights of a great scholar, colleague, and friend.

A note on the translation: Sometimes the beauty of French expression lies in allusions to subjects unfamiliar to an anglophonic audience. By adding a handful of footnotes, I have hoped to bring out some of the natural poetry without being overly didactic. The word *l'empreinte* in Philippe's original title, for example, is almost ineffable. You can read it as "footprint" or "trace," or even "fingerprint," but it has a deeper meaning of "mark" or "stamp," as of an indelible impression. Dinosaurs seem to have that effect on all of us.

The names of French organizations have been translated into English, but their initialisms reflect the original French rather than the translations: For example, the National Center for Scientific Research is abbreviated CNRS (for Centre national de la recherche scientifique).

My own thanks in this endeavor go to Jennie Dorny at Odile Jacob and especially Robin Smith, Peter-John Leone, and Michael Gnat at Cambridge, for their help and encouragement. Above all I must thank Helen Grove, Patrick Lundy, Pierre-Philippe Oriet, and Armand de Ricqlès for their best efforts in teaching me what little I know of the French language; and, of course, Philippe Taquet, for all his cooperation and collaboration, and the joy of working together on this. I would like to dedicate my labors on this book to the late Hélène Taquet, who was so greatly loved and admired by her family and friends.

Kevin Padian
Berkeley, California
December 1997

PREFACE

THE PALEONTOLOGIST EXPLORES the archives of Earth to reconstruct the history of life on our planet. In following his (or her) trade, which is also his passion, he enjoys a double privilege: Fascinated by the duration of geologic time and ensnared by the mystery of origins, he's a time traveler; as a naturalist who loves the scents of the scrub and the breath of the *harmattan*, he's a space traveler – though not yet an interplanetary one.

The paleontologist undertakes what is really a very strange profession: that of sedentary nomad as well as intellectual laborer. He's a nomad, like the Fulani shepherds of the Sahara, who say that "dust on the feet is worth more than dust on the behind." But sometimes he has to be sedentary, like the Hausa farmers of Nigeria, who say that "it takes the body's water to draw out the well's." He's the roadworker of the past, digging, picking, and chiseling to release fossil bones from their matrix. He reconstructs unknown species, bringing them back to life by reasoning and by knowledge of the laws of comparative anatomy. Sometimes it takes a lot of intellectual sweat in the laboratory to penetrate the secrets of the past.

The paleontologist has the exciting job of restoring a presence to vanished worlds from which humans were entirely absent. Baring the skeletons of strange animals before the incredulous eyes of the Tuaregs of Niger, the Berbers of the Moroccan High Atlas, or the vintners of Corbières, he is fully conscious of introducing into human thought subversive notions as to the immensity of time, our animal origins, and the evolution of species – thereby upsetting myths and beliefs, and calling into question the order of things.

In 1964, my path first crossed the trails of dinosaurs. Since then, my research has taken me from the desert of Ténéré to the *sertão* of Brazil, from the forests of Laos to the steppes of Mongolia. I've had the good fortune to discover a few dinosaurs, as well as the joy of sharing life with a great many fellow earthlings. Recounting some of these expeditions has often brought to mind the memory of a great novel, story, or film; but my aim is to draw the reader into a world that owes nothing to fiction and everything to science.

ACKNOWLEDGMENTS

MY APPRECIATION GOES to everyone who has helped me for thirty years, both in the paleontology laboratory of the National Museum of Natural History in Paris and far away during my expeditions in the field.

Without the active support of the people of the countries, far and near, where I have had the good fortune to work, without the efficient contributions of workers and researchers, technicians and administrators, I would not have been able to bring all my labors on dinosaurs to fruition.

To all of you, I dedicate this book.

CHAPTER ONE

GADOUFAOUA: IN THE SANDS OF THE TÉNÉRÉ

THERE IS A PLACE on this Earth where, simply by hopping out of your car, you risk suddenly finding yourself nose to nose with a dinosaur: a dinosaur embedded in the desert, whose spinal column, disengaged by the gritty winds from the friable sandstones that have encased it for 110 million years, is separating cleanly from the horizon between earth and sky.

At Gadoufaoua, spread over three hundred square kilometers (about 116 sq. mi.) in the south of the desert of Ténéré in Niger, there are millions of bones, dozens of dinosaur skeletons. Their diversity and quality of preservation constitute an exceptional paleontological locale, the greatest exposure of dinosaurs in Africa. It was at Gadoufaoua, in the field, that as a young beginner I learned how to study and collect dinosaurs. It was there that I experienced the dune and the *reg* for the first time.

A fortunate convergence of circumstances made me a dinosaur hunter. In December of 1964, Professor Jean-Paul Lehman, Chair of Paleontology in the National Museum of Natural History (MNHN) at Paris, invited me to pay him a visit in his office. "Invited" is the word, because no professor, no lab director, among all the talented professionals that I have ever known, was more thoughtful, more gentlemanly, more likable than he. No one else had more moral and scientific authority over the researchers in his charge. He had studied paleontology in Sweden under Eric Stensiö, the great specialist in some of the very oldest vertebrates, and was internationally renowned and revered for his learning. He also wielded a dry sense of humor with style and finesse. During his weekly lectures in his courses at the Jardin des Plantes, his excellent and often esoteric jokes made the second-year students burst out laughing, while the rest of the audience, comprising casual auditors and first-year students, sat stone-still, missing the keys to his puns and allusions.

Professor Lehman informed me with a smile that the geologists of the French Atomic Energy Commission (CEA), prospecting Niger in search of uranium, had just discovered some fossil bones. The CEA wanted a paleontologist to visit them in the field to identify these fossils, to ascertain (if possible) the age of the beds in which they were deposited, and eventually to determine the conditions of deposition and the environment that prevailed when these animals roamed that part of Africa.

My director, whose smile was becoming slightly malicious, suggested that I go to Niger. The CEA would pick up the airfare and costs. At 24, having just left the benches of the university, I was filled with the accounts of traveling naturalists, including the *Méharées* of the *parpaillot*[1] Theodore Monod, professor of ichthyology at MNHN ("the Museum," as I often call it) – whom I met frequently since we both lived on Port-Royal Square – and the *Lettres de voyage* of the Jesuit Pierre Teilhard de Chardin. I dreamed of the wide open spaces, of wonderful discoveries, and I wanted to put into practice an encyclopedic morass of theoretical knowledge that I had ingested at the Sorbonne over hundreds of hours of authoritative courses.

My immediate response to Monsieur Lehman's proposal was obviously enthusiastic. But as it turned out, I was answering yes to the following question: "Do you take paleontology as your spouse and promise to serve her faithfully for the rest of your days?" The minister of this religion, Professor Lehman, with one more malicious smile, opened his drawer, took out a plane ticket, handed it to me, and said, "Here you are. You leave in a week. Don't forget your quinine."

A week later, an Air Niger DC-6 landed in Agadès, Niger, as horsemen dressed all in white and blue pranced around the runway, preventing the livestock from crossing the stony plain that served as an airfield. At the foot of the gangway, a welcoming committee composed of geo-

---

[1] A *parpaillot* is a name ironically given to Huguenots (Protestants) in France; this word comes from Old French, and it originally denoted people who went through the streets dresssed in *chemises*, or long shirts. Monod, a truly remarkable and revered man (now over 95, as this is written), has explored much of the world, mostly on foot, in a quest for understanding of the natural world as well as for peace among its inhabitants. *Méharées*, the title of one of his famous books, is an Arabic word meaning a trip through the desert.

2

Figure 1. The Gadoufaoua locality, in the desert of the Ténéré in Niger, where dinosaur skeletons are exposed by the constant sandy winds. (Photo by P. De Latil)

logic engineers and the heads of the mission awaited the envoy from the Museum – doubtless some old professor with bifocals and white hair. They were stupefied to encounter, coming down the gangway, a student still wet behind the ears.

A little history will help here. The discovery of traces of uranium in the region of Agadès was a by-product of the study of the copper deposits that were associated with it. From 1957 to 1961, the French overseas Bureau of Mines, then the Bureau of Geological and Mineral Research (BRGM), carried out a series of prospecting expeditions west of the granite mountain range of the Aïr, located northeast of the city of Agadès.

It had been known for a very long time that there was copper in the area. In 1354 the great Arabian explorer Ibn Battûta had noted the copper mines at Takedda, a place perhaps two hundred kilometers (200 km ≈ 125 mi.) from Agadès, in the region of Azelik. Today it's an archaeological site near a mine.

In 1957, the French geologist Imreh studied this region and noticed some blocks among the sediments that had been colored yellow and green by the oxidized salts of copper and uranium. Another geologist from the BRGM, Hughes Faure, had been drawing up a geologic map of the sedimentary formations of eastern Niger since 1954. His fieldwork brought him to the southeast of the Aïr range, where he discovered numerous remains of dinosaurs; he sent several fragments to the geologist Albert-Félix de Lapparent, an exceptional and legendary man whom we shall encounter again. In a memoir on the dinosaurs of the Sahara, he noted the potential interest that this region of Niger held for paleontologists.

Since 1958, the mining prospectors' results had led the CEA to develop an important research initiative, searching for uranium all through the south and west of Aïr. Why this uranium fever? Because France, under General de Gaulle's leadership, wanted to become a great nuclear power. For that, it needed to procure the basic, indispensable ingredient of natural uranium. Now, the sources of uranium in France are mostly concentrated as veins encased in granite, principally in France's great mountain range, the Massif Central, but it was too difficult to get enough uranium out of them to form sufficient stocks. The Americans made it so that France couldn't stock up through existing producers like Canada, and this highly strategic element could not be bought legally beyond the Iron Curtain. So the French were left to fend for themselves. There were, however, important reserves of uranium outside France, in geologic contexts that were very different from the Massif Central. It was known that the great sedimentary basins of North America were very rich in uranium, but the prospecting techniques that they used there were ultrasecret and obviously not for publication. So the French geologists had to leave the Hexagon (a nickname for our roughly six-sided country), and work out for themselves the methods of prospecting, collecting, exploiting, and concentrating uranium from sedimentary rock formations. Niger satisfied all the requirements for this undertaking. So began a great technological adventure that would be crowned with success. A small team of geologists, all from the famous school of geology at Nancy, were going to have to cut corners and gain the knowledge and the necessary techniques. They had to learn in record time how to exploit one of the greatest deposits of uranium in Africa. Today Arlit, north of Agadès, is the name of a mining city and of an immense open-air quarry. The research that led to the establishment of this mine, and the construction of a town in the middle of the desert more than 6,000 km (3,729 mi.) from France, was a great technological and human adventure still largely unknown to the French people, and its story should really be written some day.

After the first moment of surprise passed, the welcome from my geological colleagues from Nancy was very warm – as was the temperature, 38 °C in the shade (100 °F). They told me that we would leave for the field at dawn the next day. A French prospector and three Niger technicians were assigned to accompany me in a Land Rover and a 4 × 4 Renault truck. Visiting the little town of Agadès – its beautiful mosque, whose adobe minaret is the tallest in West Africa, its tortuous streets and its picturesque market – would be for another day.

The next day at sunrise, we took the eastern road that led to the desert of Ténéré. Within the first kilometers, the change of scenery was complete and disorienting: the granitic relief of the mountains, the shrubby bush with its acacias, the bustards, the gazelles and ostriches, the Tuaregs and their caravans of dromedaries stretching out in long strings – everything was new to me, everything enchanted me, and I plunged ecstatically into a world whose light, contrasts, and scents I had been completely ignorant of just hours before. I had been out of Paris scarcely two days.

I couldn't help thinking of Dr. Pierre Noël, a family friend whom I had visited before leaving and who had been a doctor in the colonies in 1911. He was posted to Niger and boarded ship at Bordeaux as a young man to take up his duties at Bilma, a tiny township located on the border of Niger, beyond the sands of the Ténéré, toward Chad and Libya. It had taken him seven months to reach his post and start work, after an impressive journey: Bordeaux to Dakar by ocean liner, Dakar to Kayes (Mali) by train, Bamako to Niamey (Niger) by riverboat, Niamey to Agadès on horseback over trails. The last lap of the journey, Agadès to Bilma, was made possible by the Azalaï, the great caravan of thousands of dromedaries that traveled as far as the oasis of Djado once a year to collect dates and to bring the salt that is so necessary to man and beast from Bilma to the sedentary stockbreeders of the south. During the journey of seven months this man – who later taught a course of study for future naturalists at the Jardin des Plantes and the Musée de l'Homme in Paris – had collected sharpened flints, neolithic pottery, and insects, cared for the wounds of men and women who had suffered the attacks and pillages of rebel tribes, and drawn up a French–Kanouri grammar, all the time protecting his precious notebooks as much as possible from the ravages of termites. Seven months of traveling for him, but only 48 hours for me, while I read the paper on a couple of planes! Yet, in those earlier days the immersion into the sands of Niger was less sudden and brutal, and certainly had its own advantages.

After some hours on a trail, we left the road to Ténéré to head east-southeast, progressively navigating the strings of dunes, *regs*, and strips of very fluid sand called *fech-fech*, where our Land Rovers sank up to their wheels. With great difficulty, we crossed immense areas covered with clods of sand 50 centimeters (cm) high, crowned with tufts of tough-stalked herbaceous plants. Progress was slow and difficult; the heat and the jolts tremendously irritated my prospector guide, who favored me with his vast vocabulary of oaths.

We rolled on for 10 hours. The cars were stuck in the sand five or six times, and I quickly learned how to haul out heavy sheets of corrugated steel, place them under the wheels, and, stretching out on the ground, use a shovel or my arms to remove the burning sand – a substance that seemed more liquid than solid, because the temperature was so high. Proceeding along, we crossed many strings of dunes and rosaries of *barkhanes*, V-shaped dunes whose lee sides fell off steeply. One false move in driving meant certain disaster. Our vehicles threaded their way among these tire traps; no trail was visible. However, I learned to decipher the alignments of dromedary droppings, which occurred in great numbers on the sands. They showed the traces of camel tracks; merely by following these rather strange road signs, somewhat different from the red-and-white-painted markers of our wide European nature trails, we were able to navigate southward. This route, taken by camel-drivers for so long, wended its way astutely through piles of sand, avoiding slopes and traps. To rely on these droppings was to cross the dunes in a good mood; to deviate a few meters to the left or right meant misery, getting stuck, aggravation, and the risk of transmission damage.

Progress was very slow; our vehicles went only 10 or 20 km in an hour (6–13 mi.). And suddenly, at the bend of a long and high dune, there appeared a small mound with a wooden fork at its top. At the summit were some Tuaregs. We were at the well of Éméchédoui. It was very difficult to spot in this sea of sand; a few dozen meters to the east or west and we would have passed right by it. The few nomads present were occupied in watering their dromedaries from the well, drawing a brackish water whose purity left much to be desired. Around the well, the many skeletons bleached by the sun attested to the dryness of the region. This well had been in use for a very long time: The mound on which we found ourselves was no more than an accumulation of cow dung and dromedary droppings. This hardly encouraged us as to the quality of the well water.

We can only imagine the hypotheses and interpretations that archaeologists will form when they excavate this site 10,000 years from now. Small artificial mounds made entirely of shells, called *Kjokkenmöding,* are found in northern Europe; the Danish prehistorians who have studied them think they had something to do with kitchen debris. The site of Éméchédoui has certainly been occupied since ancient times. A few dozen meters from the well, the ground is strewn with sharpened arrowheads, potsherds, small pieces of polished stone, and even bone harpoons. Fragments of silurid fish skeletons and crocodile bones showed that we were walking on the bottom of a neolithic lake: Four or five thousand years earlier, the Sahara had been green and littered with lakes and streams. Humans and herds had been plentiful. Then the climate became dry. The presence of people, intensive deforestation, and overgrazing accelerated the process; the desert grew relentlessly, and still continues to encroach southward. An often violent seasonal wind blows during the dry season from December to March, from the northeast to the southwest. This northern wind, the *harmattan,* desiccates crops and brings enormous quantities of sand that sterilize the ground little by little. The bottoms of these neolithic lakes, lost in the middle of a field of dunes in what is a desert region today, contain the evidence of climatic changes that have come upon this part of Africa.

But at present, the purpose of our trip was to discover and study vanished worlds that were far, far older than the presence of humans on Earth. We got back in our all-terrain vehicles and set our caps eastward.

We were navigating by compass in a landscape that from far away seems absolutely flat and featureless. But up close it's a patchwork of clods of sand, *barkhanes,* sheets of gravel, and stretches of grass, without any trails and with no point of reference. Finally, toward the middle of the afternoon, a small cliff emerged on the eastern horizon. It is called Gadoufaoua, which means in Tamachek, the language of the Tua-regs, "the place where the dromedaries are afraid to descend among the bumpy rocks."

An hour later, we entered the camp of the prospecting geologists from the Atomic Energy Commission. It consisted of about a dozen tents and could be recognized only by two metallic masts visible from far away, connected by a radio antenna. Our welcome was immediate and cor-dial. The long day was nearly over; the sun plunged suddenly under the horizon, and night descended quickly. But I had a lot of trouble sleep-ing in my tent, because the impressions and images of the day's jour-ney were tumbling around so much in my head.

Early the next morning, at dawn, my geological colleagues did me the honor of showing me the locality that they had discovered. We went out by car several kilometers from there and suddenly, around a small sandstone promontory, appeared one, then two vertebral columns. But this time, they were not the bleached bones of dromedaries. The verte-brae were of a somber color, bluish and stony. Some of them were still encased in the sandstones that had protected them from destruction for millions of years. The spectacle was striking. Imagine . . . no. It's unimag-inable, because there is no equivalent in nature. Over many hundreds of square meters, huge skeletons were embedded in the sand, lounging or lying on their flanks like dromedaries at night. The undulations of the sand reminded me of a beach where a herd of elephant seals were stretched. But these were not dromedaries, nor elephant seals: They were dinosaurs.

Leaping from the car, my heart pounding, I headed toward the best preserved among them. I had the exhilarating feeling of being in the skin of Professor Challenger in Conan Doyle's famous novel *The Lost World* as he discovered a herd of living iguanodons in the heart of the Amazonian forest.

A first glance told me that the the bones were superbly fossilized; they were solid and well mineralized, and the structure of the bony tissues was perfectly preserved. These skeletons had not been disturbed by ero-sion or by movements of the Earth. The gritty wind had worked like sandpaper or any other abrasive, naturally disengaging the fossil bones bit by bit. The result was striking and unique: The vertebral columns of these dinosaurs crowned the summits of little mounds and were sep-arating themselves along the horizon between sand and sky. The two best skeletons were seven or eight meters long (23–26 ft.). It was easy

Figure 2. The complete skeleton of an *Ouranosaurus*, a herbivorous bipedal dinosaur, lying on its side like that for 110 million years. (Photo by P. Taquet)

enough to recognize the front ends of these great reptiles: The neck vertebrae were broad disks, convex in front and concave behind; those of the tail were quadrangular and their size decreased progressively toward the end of the tail, which was still recognizable. Between the neck and tail vertebrae, the dorsals were perfectly visible: They were more massive, and flanked by two sets of flattened ribs. The death-poses of the skeletons could be seen easily on the ground. These nearly complete dinosaurs, separated from their rock matrix in natural pose and stretched out on the sand in a place so deserted and quiet, made such an impression that every one of the assembled scientists fell off talking. The dinosaurs of Gadoufaoua, more than a hundred million years after their death, had the right to a minute of silence.

We regretfully left this unique spot, which the CEA geologists had christened the Cemetery of the Innocents.[2] All through the day we went from place to place finding more bones. From the abundance of fossilized remains on the ground, their diversity, and the quality of their preservation, I could see right away that the Gadoufaoua area was exceptional. The following days' explorations reinforced that impression and made me realize that this was one of the most important dinosaur-bearing sites in Africa. As it turned out, the strip of land that yielded all these bony remains is about 2 km × 150 km – about a mile wide and a hundred miles long! This was almost unbelievable: it required a 110 million (plus 1,964) years for this discovery and a favorable confluence of circumstances before this January day, when I found myself, a lone fledgling paleontologist, in the middle of an immense dinosaur cemetery. Armed with my geologist's hammer, my notebook, and my pencil, I shouldered a tough new burden of responsibility, as a wave of elation swept over me from my toes to my scalp.

The geologists looking for uranium needed to know the age of these fossil beds. The study of these bones and the paleontologist's determination of what they were would answer their questions; that's why I was invited. But before I could see this part of the research through to its conclusion, I needed their fieldwork to allow me to understand the lay

---

[2] The "Cemetery of the Innocents" is where Parisians buried their dead until 1786, when a flood of the River Seine hydraulically exhumed the ageless corpses and floated their putrid remains through the Paris streets. When the waters receded the dead were reburied in the Catacombs, which had been Roman quarries, at what were then the city limits; the remains can still be seen, but the spectacle is not for the fainthearted. The French Resistance was headquartered there during World War II.

of the land, as well as its position in relation to other geologic beds. That long band of rock rich in dinosaur bones represented the surface exposure of *one bed*. A few words of explanation on their part was enough for me to understand the importance of the geometry of the rock strata. A hundred and ten million years ago, this region was bordered on the south by some great mountains in Nigeria. Broad rivers, at first torrential, ran from south to north; their great meanders then stretched out along what was then an immense plain strewn with marshes and lakes. On this humid plain, where a hot climate prevailed, conditions for life were particularly favorable for a whole fauna of dinosaurs, crocodiles, turtles, and fishes. Here, over the course of thousands of years, these rivers desposited enormous quantities of detrital sediments: coarse and fine sands, clays, and silts. In this way a range of horizontal sedimentary beds was laid down in a vast basin, in the middle of which were entombed dozens of skeletons and millions of fossil bones. After these beds were deposited, they were overlaid by other kinds of beds, which eventually masked the levels that were so rich in the remains of vanished animals. The ocean then made a short incursion on the Saharan platform, leaving limestone beds and shells that showed they were deposited during the Cenomanian stage, in the middle of the Cretaceous period – so the Gadoufaoua beds were obviously earlier, dating from the Early Cretaceous.

If things had stayed like that, we would have had to dig a mine shaft to get at the fossils today, buried as they were under all those younger rock beds. Fortunately, a geologic event intervened: The mountains of the Aïr, located north of this sedimentary basin, were uplifted, and in turn they tilted the beds, which were originally horizontal. Because of this, part of the fossiliferous layer reappeared at the surface, and the erosion caused three thousand years ago by rains, and today by the sandy winds, stripped and wore down the sandstones, naturally freeing the bones that had been trapped. But only the edge of this basin was elevated, so the original sequence of the beds is known only from their edges. You can think of this basin as a book whose pages represent the geological beds, and whose edge is the only part of the beds that can be seen today. Open the book slightly and you'll see a thin slice of the outer margins, a little distorted from their closed position, appearing at the surface. This demonstrates why the dinosaur-bearing level that interests us is in such a narrow band, a hundred miles long and only a mile wide.

Through the course of geologic time, the sandy sediments were transformed into the fairly solid sandstone that forms the relief of this little
cliff. It has resisted erosion better than the clay beds below and immediately above it. This cliff is clearly visible in the immense desert, and the
dromedaries hesitate to cross it because they're more used to the flat
and sandy terrain. This is why the Tuaregs call it Gadoufaoua.

For three weeks, as the only paleontologist in the heart of this magnificent desert, I could devote myself to one of the most exciting jobs of
my trade: prospecting and looking for fossil remains. These could be isolated teeth or jaw fragments, limb bones or vertebrae, ornamented bony
plates or fragments of carapaces, shells or pieces of silicified wood, scales
or phalanges. But there was always the possibility of discovering skeletal elements, or even better, complete skeletons. People often think that
this is a matter of luck, but it isn't. Actually, to come up with good finds,
certain conditions have to be met. First is obviously to find the right geologic level, and for that you have to understand the geometry of the
different deposits. When you follow a fossiliferous outcrop along the
ground, at the same time you're figuring out its thickness, its inclina-
tion (or, as we say in our jargon, its *dip*), its orientation (north–south
or east–west, for example), the events that might have overturned it,
the faults that have fragmented it, the movements of the Earth's crust
that might have crumpled it, and the recent deposits, such as sand dunes
or lake beds, that later came to hide it in places.

The second condition is tied to another type of activity: walking. To
find fossils, you have to walk. Walk a lot. For hour after hour, and day
after day. You have to survey the ground in every sense, methodically.
It's like being a mushroom hunter, looking for those prized and exquis-
ite *cèpes* or chanterelles among the mosses and rotting bark in a damp
forest. You have to use your good legs and your sharp eyes in all conditions: in the middle of the desert, under a scorching sun or a burning
desert wind, in the mugginess of a tropical glade, with your feet in the
water, or on the steppe surrounded by a violent glacial wind. No matter how tired you feel, you have to keep moving, and looking.

The third condition has to do with the profession of the naturalist,
of the paleontologist. Where the ordinary person out for a walk would
pass right by a bird, the ornithologist recognizes the warbler or bull-
finch. Where the traveler would only see stones of various shapes and
colors, the fossil hunter sees fragments of silicified wood or the ends of
limb bones. Above all, thanks to training in anatomy, he (or she) can

recognize condyles and apophyses, sutures and trochanters – those details of biological form. The structure of a piece of fossilized wood or bone can be recognized easily, if you know what you're looking for. And then, in the field, it helps if you can pick out, from the middle of an area strewn with bone, those bony fragments that are worth talking about, the ones that can give you information about the animal to which they belonged: its anatomy, its way of life, the conditions of its death and fossilization. In fact, any paleontologist in the field is constantly making subjective choices. Some collect many fragments, not all of interest, whereas others have the gift of finding and collecting pieces of the skull or even complete crania, new skeletal fragments belonging to species unknown to science. Sometimes you never know until you get it back to the lab.

Finally, to make great discoveries, you also need a kind of intangible skill that often seems like flair, experience, or luck. You can pass several meters by an exceptional specimen without seeing it, but you can also stumble on the end of the tail of a dinosaur barely sticking out of the rock. This is how you get into the second phase of paleontological research: extracting the most interesting fossils from their matrix.

So the bone hunter has to be a successful geologist, rambler, and naturalist. And then he has to transform himself into a roadworker, a sculptor, a plasterer, and a trucker.

At Gadoufaoua, there were so many fossils at this new locality that my first prospecting task consisted of mapping the most favorable zones and then, within these zones, selecting the most complete and least eroded dinosaur skeletons. One fine morning in the zone north of the locality, south of an abandoned well called El Rhaz, I found two fairly complete dinosaur skeletons stretched out on the ground about 50 meters apart. An excavation at that spot, I thought, ought to bring results that would help both paleontology and geology. So I took some sightings, built two small piles of stones that could be seen from far away, took some photos, made a sketch of the setting, and jotted down a preliminary account of the bones exposed on the surface. Those were my first dinosaurs. I knew that the discovery was exceptional, and that the locality would become famous; but I was the lone paleontologist in this desert place, armed only with my geologist's hammer and some plaster bandages. I would have to come back with a whole crew, with all the necessary materials and enough time to do it right. It was on that day in January 1965 that I became a dinosaur hunter.

Some weeks later, carrying a few kilograms of fossils in my bags as spectacular evidence of this vanished world frozen in the middle of the Ténéré desert, I headed home, blinded by the African light, seduced by the beauty of the desert, and carried away by so many discoveries.

I brought the fruit of my labors back to Paris, but I had also discovered in Niger a country very different from my own, and people different from my compatriots – warm, exuberant, gay, and generous, despite their extreme poverty. My return on a gray February day to the City of Light, with its Métro full of sad crowds, men and women with stony faces, was a bit difficult. After an enthusiastic, passionate account of my mission, I threw myself into organizing an important expedition that would return as soon as possible – as soon as the next dry season came – to Gadoufaoua, back among the herd of dinosaurs that awaited me.

The CNRS (National Center for Scientific Research) recruited me as a trainee a few months later. My initial salary was clearly not enough to finance such an operation (a CNRS research trainee earned 800 francs a month, or about $160 or £67 at that time), but my induction into this prestigious research organization allowed me to get official approval and funds to go in the field. My director, Professor Lehman, added the influence of the MNHN. The support of these institutions made it possible to sign an official accord with the Niger authorities, particularly with its Minister of Education. One of my colleagues in the Museum, Léonard Ginsburg, next to whom, as a young student, I learned the rudiments of fieldwork, agreed to go with me. A student, Bernard Loiret, joined us. Then we had to find all-terrain field vehicles. The Center for Research on Arid Zones, based in Algeria, loaned me two Land Rovers and an excellent driver and mechanic, Michel Séguy, whose role would be crucial. You see, a paleontological expedition is really not a Hollywood-style adventure or an intrepid and danger-filled, Indiana Jones–type of raid. An expedition made up only of passionate bone hunters, unfamiliar with the desert, would risk losing their possessions and lives on the first trip. A paleontologist might bring with him bags for collecting specimens and digging tools, but he might forget spare gas cans and a repair kit. And that could have really unfortunate, even drastic consequences. Just because you can put a skeleton together doesn't mean you can reassemble a motor from its pieces.

The months of preparation rolled by quickly, and in early 1966 we were ready to head south. I was burning with impatience to get back on the trail to Gadoufaoua.

On February 25, 1966, our little crew took the plane to Bourget Airport and reached Algiers to hook up with Michel Séguy and his two vehicles, brought by boat. A long trip by road, then along a trail, took us to Agadès via Laghouat, Ghardaïa, El Golea, In Salah, and Tamanrasset. This was the opportunity of a lifetime: to initiate myself into the geology of the African continent by crossing its ragged northern border, then its horizontal platform on which three hundred million years of continental sediment had been deposited from the south – since the Carboniferous period, represented by the mountainous jetty of the Hoggar. All the phases of the history of a continent, from the most recent to the most ancient, succeeded each other from north to south. For the geologist, all of these often imposing lands, and all of their sometimes spectacular relief, have a history and an explanation. Like the tourist, he appreciates the spectacle of nature, but beyond this his trade allows him to appreciate the origin of each of the surrounding elements. The hiker admires the mountains, powerful and immovable; the geologist sees them born, rising, folding, aging, and disappearing.

Five thousand kilometers by road, then by trail to reach Agadès and then the dinosaurs of Gadoufaoua – an excellent initiation to all-terrain driving; it would be tremendously useful to us when we continued our fieldwork without the aid of guides and drivers from the Atomic Energy Commission. The year before, I had taken as many sightings as I could to retrace our path under the best conditions. It wasn't easy, but after many hours of navigating a sea of sand, we reached the end of our journey.

My two dinosaurs were still there, and apart from the occasional visits by ostriches, oryx, addax, and gazelles, the places had remained deserted. We set up our camp at the foot of two acacias, in a tiny valley that protected us a little from the wind – as if these trees, the only ones that could be seen for miles around, provided a refuge, a more welcoming place amid the immensity. In any case, they served as markers for the whole time that we worked in that area.

Early next day, we started our work, collecting elements from the first of the two dinosaur skeletons located in the previous year. In this place where the sandy wind had already freed many bones, the first task was to brush away the sand that had recently covered part of the surface. All the bones and all the bony fragments sticking out of the ground were cleaned but left as we found them. Then the excavation proper began: With small trowels, nicknamed "cat's tongues," all the superficial, friable sediment was removed. Under this eroded coating, the sediment became harder; then we had to use hammers and chisels. With great care, using precise and measured strokes, we detached blocks and fragments. Bit by bit, the fossiliferous exposure was cut into small pieces, but each bone had to stay in place, encased in its rock matrix. With unwavering attention, brushing the surface constantly, we worked without touching the bones or risking the loss of some interesting piece. Little by little, we revealed the parts of the skeleton. The tail vertebrae, with their square bodies and short apophyses (spines), were connected, still associated in the order that they occupied in the living animal. Some meters away, two parallel long bones were now visible, one fairly massive, the other a little slighter; their forms allowed us to recognize them as the lower leg bones (in our jargon, the tibia and fibula). Still farther off, some long, flattened, and recurved bony bars were protruding from the ground – these were ribs; while there on the ground lay a fragment of the upper part of the skull, the skull roof, easily recognized by its sutures, the jagged lines that join two bones. In the space of a few hours of feverish but precise work, we realized that this dinosaur, discovered the previous year, was relatively complete. Gradually it revealed itself to us, lying on its left side, its tail stretched out and its hindlimbs and rib cage clearly visible.

We needed about 12 days of unremitting work to uncover the whole skeleton. One of the most exciting tasks in paleontology is to bring a dinosaur skeleton out of its cocoon of rock. This one measured eight meters in length, and it was so perfectly preserved that we could see some important skeletal features. Its dorsal vertebrae were topped by very long apophyses, or "neural spines." The length of these spines was surprising: For a vertebra whose body, or "spool," was 16 cm tall ($6\frac{1}{3}$ in.), the spine was 60 cm (~2 ft.)! This dinosaur, with its high-spined backbone, must have been quite something when seen from the side. At the level of the animal's pelvis, two large symmetrical bones proved to be the left and right pubes. They were formed in front by a great flattened plate with rounded borders, and behind by a long, fine rod of bone. This type of pubis is very characteristic of the bipedal, herbivorous ornithopod dinosaurs called iguanodontids and hadrosaurs. Eventually, near the neck vertebrae, we found one of the lower jaw bones of the animal. This is called the *dentary,* and above it, we found some well-preserved spatulate teeth, lance-shaped, with one side covered with enamel and the other scalloped. They reminded me of the teeth of *Iguanodon,* the

Figure 3. The dorsal vertebrae of *Ouranosaurus* had long neural spines that could reach a length of 60 cm (23.6 in.). (Photo by P. Taquet)

dinosaur known at first only from teeth found in England by Gideon Mantell, a country doctor in Sussex, and his wife, in 1824. Was this an *Iguanodon* or something different? It was much too early to tell, but what was certain was that this was an exceptional discovery: For the first time, a French paleontological crew had the opportunity to collect a complete dinosaur.

For two weeks, working seven hours a day under a broiling sun, we pursued our task. We now saw that the animal's neck had been reflexed over its back. Its posture was like those of the dromedaries and donkeys that die at the edges of the trails in Niger. As when mammals die, the contraction of muscles on the upper side of the neck drag the head backward; what we see in carcasses today is exactly what we saw in the dinosaur skeletons. This one was quickly buried in a fine sandy sediment after its death. The bones of the skeleton remained in articulation; they hadn't moved, and they were fossilized that way. The dinosaur had been locked in its death position.

Most of the bones of the skeleton had been excavated with the greatest care. Now what we had to do was to impregnate the most fragile bones with liquid glues. We had brought these from France; you can make them more fluid by adding solvent, such as alcohol or acetone, which helps them penetrate the bones. Then they evaporate, which happens sooner when the air is drier, as it was where we were. The bones, hardened in this way, can travel well without breaking, whereas the pieces with fresh breaks eventually can be assembled without much trouble. These glues are also great because they can be dissolved if you want to examine the internal structure of a bone, or if the bone fragments have been been poorly assembled.

Once we finished this assembly work, we had to record each piece on graph paper that charted its position on the ground, so that the numbers would correspond. We used a very simple code; for example, GDF 300-14, in which "GDF" referred to the locality, 300 was the specimen number (in this case, a bipedal dinosaur), and 14 was the number of the bone of that skeleton (for example, a left humerus). If a bone was found broken into pieces, as was often the case, each fragment had the same number. This task, though long, was indispensable: It allowed us to relocate the original position of each bone; that let us reconstruct the different parts of the skeleton and often avoided errors in interpreting certain anatomical features. For example, the painstaking excavation of this dinosaur specimen and its reference points allowed me to show that the bony bar that forms the front part of the right pubis was narrowly connected to the left pubis. This arrangement had never been observed before in ornithopod dinosaurs.

Mapping specimens is an exercise in geometry or specialized cartography that is usually pretty easy in France and in temperate climates. But in Niger, it could take on acrobatic proportions once the sandy wind started to blow. If you can imagine how hard it could be to use a drawing board in a wind tunnel, I can tell you from experience that it's worse in the Ténéré.

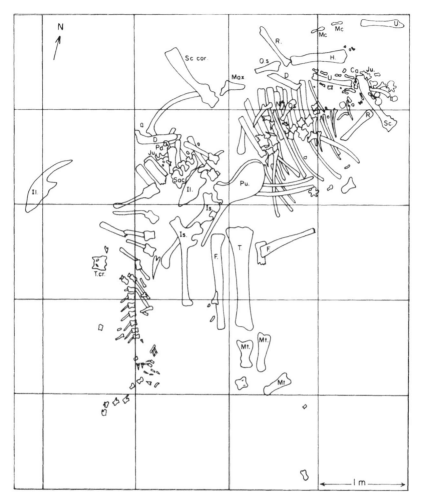

Figure 4. *Ouranosaurus nigeriensis,* a new genus and species of dinosaur. This is how the type specimen was found in its quarry. (Drawing by F. Pilard)

The painstaking inventory of every piece on the outcrop was the kind of job you had to do over a long spell because, if the skeleton was complete, as it was here, you had to expect to find a respectable number of bones. But how many? A quick inventory gave us 80 vertebrae (neck, back hip, tail), 27 pairs of neck and back ribs, maybe 54 bones; 21 bones from each forelimb, about 42 for the two forelimbs together; 24 for each hindlimb, maybe 48 bones; 10 for the girdles, that is to say the shoulders and hips; and finally about 40 for the skull. So that made a total of 274 bones, each broken into several pieces; overall, some 600 fragments! Each one had to find its rightful place, through a series of operations starting with collection, then packaging, then transport, preparation, and finally reconstruction in the laboratory – hoping for an animal that wasn't a chimera,[3] but rather a faithful reconstruction of an animal that no one had ever seen, one that populated this part of Africa 110 million years ago.

The reference points of the terrain established, we started collecting. All the little isolated pieces were carefully wrapped in paper – safely cushioned. Each bony fragment and each complete bone was wrapped separately to avoid rubbing and wear during transport.

But the most delicate work remained. With the help of hammers and chisels, we set to work cutting away the various kinds of rock that had encrusted the skeleton. To do this, we had to hollow out a series of deep and narrow grooves around the concentrations of bone without touching or damaging the pieces that interested us. A paleontologist is rather like a quarryman, a roadworker, and a silversmith all rolled into one. You have to break rock, sometimes with massive strokes of a sledgehammer, and you have to shovel and dig, but you must also constantly control each movement, to avoid damaging the pieces and setting off the vibrations that will fracture the bony remains. Hence the skeleton of our dinosaur was cut up into a number of blocks whose bases were still stuck solidly in the ground. Each block was then encased in a solid jacket of plaster. To do this, first you have to protect the surface bones by covering them with wet paper. Then, just as the surgeon plasters a patient's broken leg to hold the fractured bones together, the paleontologist protects the blocks and bones with strips of burlap soaked in wet plaster. After these dry, a very solid jacket covers each block, protecting the precious bones from external shock while maintaining the necessary cohesion on the inside. The operation that follows is relatively delicate. You have to slide crowbars, flat chisels, or levers under the base of the block to detach the jacket from the underlying rock mass. Once the block is detached, you have to turn it over and embed the unplastered side with strips of plastered burlap.

---

[3] A chimera is an animal formed of the parts of many separate animals (in Greek mythology, a monster killed by Bellerophon). Paleontologists have to be careful, in excavating a fossil locality, not to associate pieces of different animals in the same skeleton, thus unwittingly producing a "chimera."

This method, although extremely simple, is remarkably practical and was used early on by bone hunters in both Europe and North America. It lets you dig out and carry away a large number of fossils in a relatively short time in very good condition, and to transport them to a suitable place so that you can then have all the time you need to excavate and prepare each bone in the laboratory.

At the end of two weeks of exhausting but strangely absorbing work, a dozen plaster blocks, ranging from 25 to 50 kilograms (kg) (55–110 lb.), as well as an imposing number of smaller jackets, contained everything that we had been able to collect – everything on the ground that belonged to this splendid skeleton.

All of our activity obviously took place at the mercy of the sun and the wind that never stopped blowing the whole day, raising clouds of sand. This sand penetrated absolutely everything, "even inside the eggs" as the Tuareg proverb goes, and I could easily believe it. It irritated the eyes, and on some days the wind was so powerful that it was filling our excavations almost as fast as we were digging them out. Apart from that, we had quickly adapted to the heat and dryness of the region. The great thermal leaps between night and midday exposed us to temperatures varying from 3–4 °C to 40 °C (37–39 °F to 104 °F) and 4 percent humidity. We had adopted the turban, the *chèche* of the Tuaregs, which protected us perfectly from the sun and the sandy wind, while a good pair of traditional leather sandals, called *samaras,* allowed us to walk on the soft sand with ease.

Eventually we became accustomed to the fairly rudimentary conditions of life in the desert. As the days went by, the strangeness of the country and the solitude of the sites led us, little by little, to enjoy the charm of these privileged moments. What surprised the city-dweller that I had become in Paris was the absolute silence – a silence that you discover, that you appreciate, that you hear.

You could *hear* the silence.

In the legends of desert peoples, a large, open space is said to be rife with the whisperings of genies and spirits. This is no surprise to anyone who has spent time in the desert. The simple passing of a wisp of sand on the dune, the nightly emergence of a jerboa near the tent, were perfectly audible in the great silence, which was barely broken by the occasional passage, high in the sky, of a mail plane. And then there were some nights, under a spectacular sky, when all the constellations and the Milky Way seemed to be at arm's reach, while the Southern Cross, so beautiful and fascinating, was surrounded by fireworks of shooting stars. Time, sand, and distance take on entirely new meanings in the desert.

One beautiful morning, three dark points appeared on the horizon. At first tiny, they soon enlarged to transform themselves into silhouettes that we initially took for three frolicking iguanodons – a normal enough mirage under the circumstances. Work stopped on the site. As in fairy tales, and no doubt African legends, these three strange creatures soon changed into three dromedaries, each carrying a magnificent Tuareg. Dismounting a few meters from us, these three men, dressed entirely in blue, silently approached the skeleton embedded in the rock, which we were in the process of excavating. Their surprise at our work instantly galvanized the camp. What would they make of all this? How did this curious giant dromedary, this monstrous lizard, end up in the middle of their desert? In the nomadic memory, nothing like it had ever been seen up to now in the entire Ténéré. Besides, doesn't the word "Ténéré" mean exactly that – "nothing"? There was supposed to be nothing but sand in this immense desert region.

The Tuaregs scrutinized our find carefully, then moved away. I have always remembered the image of these three reserved but perplexed men, so noble in their demeanor, commenting in their Tamachek language on this strange apparition in the middle of their desert.

You can't meet up with the nomads of Niger without the tea ceremony. Three cups per person; three cups of a very hot, syrupy, and strong brew. Our local guides asked where they came from. The eldest of the trio, stretching his arm northwest, answered: Agadès. Where are you going? The same man, sweeping his arm southeast, said: Termit. All in all, a trip of 400 km (250 mi.) across the desert, in a straight line or nearly so, without a trail, without a map, and without a compass. But these great travelers had the quiet assurance of arriving right at their destination: a well, hidden at the base of a small cliff far, far away from here. A few words of goodbye and the trio set off, disappearing quickly beyond the horizon, leaving us in silence. For our part, we had certainly been more impressed by their passage, sort of a walking mirage, than they had been struck by our presence in these parts and our robbing of antediluvian graves. This unusual encounter between Tuaregs, who measure time by a day's travel, and paleontologists, whose concept of time is measured in millions of years, was surely unexpected.

But doesn't it bring home the relativity of time?

With the first skeleton collected, we turned our attention to the second skeleton located the previous year, a few dozen meters away. The second specimen soon turned out to be very different from the first: stockier, heavier, more massive. Its forelimbs were as powerful as its hindlimbs; we were certainly not dealing with a bipedal animal, but a quadrupedal one. However, the body, or "spool," of its vertebrae had the same squared-off shape that you find in iguanodontids. This shape was especially characteristic of *Iguanodon bernissartensis*, of which a troop of 31 skeletons had been found in 1878 at the bottom of a coal mine in Bernissart, Belgium. Preparing this skeleton in the laboratory would later enable me to understand the relationships between this African iguanodontid and our European ones.

A new week of work and our second dinosaur was in its turn excavated, plastered, and wrapped. We still had a few days at Gadoufaoua to prospect the extensive fossiliferous exposures that surrounded our camp, and very soon the discoveries multiplied, so abundant were the fossils at that site. A small sandy mound yielded several small turtle shells; a second held the remains of a small short-snouted crocodile; a third, formed of great sandstone slabs, held flattened bones and jaw plates, armed with a multitude of rounded, buttonlike teeth, belonging to a large fish that probably crushed shells. Then, a considerable surprise: On the ground, largely prepared out by the sandy wind, lay the gigantic skull of a crocodile with an extraordinarily long snout, much like the fish eater of India called the gavial. This skull was 166 cm (~5$\frac{1}{2}$ ft.) long, and bore enormous conical teeth set into cylindrical sockets. We shall have more to say about this monster later on (in Chapter 4). Our fossil collecting was fruitful, and we had an embarrassment of riches from which we had to choose among the millions of bones and scattered fragments: sharp, pointed teeth of carnivorous dinosaurs; slender, barbed, needlelike shark teeth; flat lungfish teeth, ornamented with large curving crests; neck vertebrae from quadrupedal herbivorous dinosaurs; sharp, curved claws of carnivorous dinosaurs; thick plates of the sculptured dermal armor of crocodiles; great shells of curious freshwater mussels still planted upright in the sediment, forming strange bouquets that allowed us to visualize the course of the river that had flowed through here 110 million years ago. And to top off this evocation of a vanished world with its abundant fauna, the silicified remains of a luxurious flora, enormous tree trunks set end to end, stretching 15–20 meters long (50–65 ft.), and lumpy, foamlike piles of plant structures. We really had an embarrassment of riches; we were the first to collect fossils in this part of the world, and the wind's erosion had, for hundreds of dozens and dozens of hundreds of years, prepared and extracted for us all these pieces, all these paleontological treasures. Our final prospecting had shown me only that the abundance of fossils was every bit as great elsewhere along the immense strip of outcrop of these beds. I would soon understand that my first paleontological mission had brought me to the largest and richest exposure of dinosaurs in Africa.

Figure 5. *Ouranosaurus nigeriensis:* reconstruction of the skeleton, 7 meters long (23 ft.). (Drawing by F. Pilard)

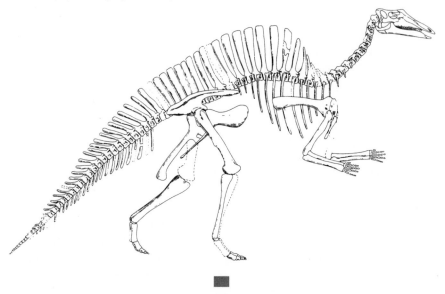

■

CHAPTER TWO

THE *OURANOSAURUS;* OR, HOW TO BRING A DINOSAUR BACK TO LIFE

T O TRANSPORT A DINOSAUR, even in separate pieces, is not a trivial affair. When the transport is between the Ténéré and the Jardin des Plantes in Paris, the compounded difficulties make this operation particularly delicate. The first problem was to pack the specimens so that they could resist the shocks of the trip, by using good strong planks of wood. But in Niger, where two-thirds of the territory is desert, most of the available wood is silicified – fossilized. So our only solution was to take the used but still sturdy crates that had brought material from France for the uranium prospectors. Our two dinosaurs were accordingly boxed up and transported to Agadès on a strong all-terrain truck. But then we had to go through the usual administrative procedures to get the necessary authorizations to export this unusual merchandise. That took some time. Finally, we had to find a safe way to transfer the material from Agadès to Paris. The solution turned out to be the French civil aviation planes that were bringing airport equipment to Africa and returning empty. This also took time. When transportation was arranged some months later, I discovered with horror that our beautiful and precious crates had become termite food, reduced to ruins and dust! We had to do it all over again: find wood, rebuild the crates, repack our dinosaurs, and send them as soon as we could. We found an elegant solution in the reuse of something the mining geologists often used in their work: long crates for holding the cylindrical cores of rock they had pulled from the subsoil with their drilling tools, which they called "carrots" because of their tapering, pointed ends. These long boxes were perfectly adapted to hold the limbs and long spines of our dinosaurs. Soon our load was ready to go once again.

■

It was no simple task to hoist our precious load as high as the hold of an airplane on the runway of the Agadès airport without equipment. A truck with a forklift and winch helped us to lift the crates one by one and push them into the cabin with considerable difficulty. I had repeated visions of each one in turn falling and crashing to the ground. To watch helplessly while a tibia or femur of our dinosaur shattered would have been a disaster, since all our efforts had been dedicated to bring our patient safe and sound back to the doors of one of the best clinics specializing in animal fossils, the Institute of Paleontology at the National Museum of Natural History in Paris.

Despite these fears, the boarding went without a hitch, and with relief we watched the plane take off, carrying in its flanks the giant reptiles that never could have flown by themselves. We would catch up with our collections a few weeks later, on our return to France. But the arrival of our material at Bourget Airport and the customs operations that followed turned into a comedy of errors. You see, our crates still bore the miners' label, "drilling carrots," so the customs officials asked us to justify why we had to bring tons of carrots into France. The administrative boondoggle was quickly cleared up: Our carrots were not edible, and in any case our dinosaur bones would not be competing with the produce of French truck farmers.

Once our special cargo arrived in good condition at port, that is, our laboratory, a new phase of my research could begin – a phase that would last a whole year, dedicated to preparing, cleaning, and extracting the fossil bones collected at Gadoufaoua from their matrix.

This manual labor requires patience and delicacy. With small chisels, and then with smaller electric pick-hammers bearing fine needles with solid steel points, each piece was disengaged little by little from the sediment that surrounded it. The tiniest bony fragments were put back in place, and the fragments of a single bone were reassociated and reassembled. Experience made the task easier: We quickly learned to flake off the rock by sliding in a chisel parallel to the bedding plane, stopping the taps of the hammer before it reached the bone, and using the natural surface of discontinuity between the bone and the matrix. In this way, it was easier to detach each portion of sediment that clung narrowly to the bone. With the Gadoufaoua sediment, mechanical preparation of the fossils was the only possible method, because the rock was sandy and siliceous. Had the matrix been calcareous, we could have prepared the specimens chemically, plunging the fossils into a vat of a dilute solution of acetic acid to extract the denser bones by dissolving the rock. The sediments of calcium carbonate dissolve more rapidly than the bone, which is made of calcium phosphate.

Today, technical advances let us remove bones from a sandstone matrix using jets of compressed air shooting abrasive powder or sand through a very fine tube. This very efficient tool, developed by industry to smooth metallic pieces, has been widely adopted by paleontologists. In 1966, though, it wasn't available to us, so most bones collected at Gadoufaoua were cleaned by hand, one at a time, by the preparators in the Museum's paleontology laboratory, who are especially skilled and resourceful technicians.

The prepared bones were assembled little by little; their most fragile parts, which were spongier and risked crumbling, were impregnated with thin liquid glue. The most solid portions were left as they were. Then each piece regained the number that had identified it in the field (GDF 300-25, GDF 300-26, etc.).

As the days passed, all the elements that made up our skeleton – at least, all those that had been preserved and that we had been able to collect – were prepared and repaired. Then began the exciting work of identifying and interpreting each bone. Manual labor was succeeded by intellectual labor. Today, each paleontologist constantly applies the rigorous methods and approaches of that brilliant promoter of comparative anatomy and vertebrate paleontology, Georges Cuvier, who in 1812 wrote at the beginning of his *Researches on Fossil Bones:*

As a new kind of antiquary, I have had to learn to decipher and restore these monuments, to recognize and bring together in their original order the scattered and mutilated fragments that constituted them; to reconstruct the ancient beings to which these fragments belonged; to reproduce them with their proportions and their characters; eventually to compare them to those living today on the surface of the globe: a nearly unknown art, that presupposed a science that was hardly developed in former days – that of the laws that govern the coexistence of the forms of different parts in organic beings.

After learning to restore the "monument" that was this dinosaur skeleton from Africa, I now had to learn to recognize its scattered and mutilated fragments and reassemble them in their original order. It was relatively easy to put a name to many very characteristic bones. The femur or thigh bone was instantly recognizable, with its slightly oblique articular head and its lower end, equipped with two large condyles, or rounded surfaces, against which the upper end of the tibia (shin bone) articulated. So was the humerus or upper arm bone, which had a broad, palette-shaped upper end with a thick external border, or deltoid crest, that anchored the muscles that raised the arms. It was equally easy to distinguish a neck vertebra from a tail vertebra: The first was light and made of fragile arches, whereas the second was massive, with a squared-off body. But it was much more difficult to put a name on the thin bony shards of the palate, or to retrieve the exact order of the long bony rods (the metacarpals) and the rounded bones (the carpals) that comprise the skeleton of the hand.

In each case, the method was the same: to examine attentively, "under all its seams" as we say, each indeterminate bone; to pore over its articular facets, their forms and orientations; to read the presence of the sinuous suture lines that join two bones of the skull; to look for possible contact surfaces, observe the rough traces of muscle insertions, and take inventory of the different foramina, the small holes in the bone wall that show the passage of blood vessels or nerves. This is how we figured out the exact place of each bone among the some 274 elements that constituted this skeleton.

Cuvier has generally received the credit for clearly showing that living beings are constructed along different plans of organization. These differences in construction allowed him to group animals into what he called *embranchements:* vertebrates, molluscs, articulates, and zoophytes (or radiates). Crocodiles are vertebrates, oysters are molluscs, insects are articulates, and sea urchins are radiates. Since that time, other *embranchements* have been defined and argued, but these distinctions have allowed scientists to study each *embranchement* for structures that could be identified as homologous.

The various principles of construction of living beings, starting with those that the great anatomists of the nineteenth century discovered for the vertebrates, were very useful tools and highly efficient for giving us a name for each of our dinosaur bones, for finding each one's place amid the architecture of the skeleton, and for understanding the relations of each one with its neighbors.

The body of all vertebrates is obviously organized along an axis, the vertebral column, with an anterior extremity (the head) and a posterior extremity (the tail), and it also has a symmetry of two equal halves, the left and right. Because the vertebrate body has this bilateral symmetry, we can distinguish paired from unpaired structures with ease. Ears are paired, the nose is unpaired. In the same way, the skeleton has paired and unpaired bones. A vertebra is unpaired, and its architecture reflects its position along the midline of the body: one side of it is symmetrical with (i.e., a mirror image of) the other. But the radius and the tibia are paired bones: There is a right radius and a left radius that are symmetrical with each other. The skull also has paired and unpaired bones: the frontals, lacrimals, and dentaries are paired, but the predentary and the reptilian basioccipital are unpaired.

Although the distinction between paired and unpaired bones is simple, it can be hard to determine some bones when they come from extinct animals. The methods of determining fossil bones have developed over the centuries along with the natural sciences. A Danish anatomist, Nicolas Steno, who practiced in 1667 in Florence, gets the credit for being the first to show how to apply sound reasoning to such determinations. While studying fossils found in the sandy sediments on the outskirts of Florence, Steno showed that the objects that had been described as *glossopetri* (or "tongue-stones") were identical to the teeth of large living sharks. On the basis of these teeth alone, he was able to deduce that sharks had lived in ancient times and that their skeletons were similar to those of sharks that populate the oceans today. This comparative method that Steno devised was developed and formalized a century later by Cuvier. To Steno's comparative method Cuvier added the use of the *principle of the correlation of parts,* which he described in this way:

Every organized being forms an ensemble, a unique and closed system, of which the parts mutually correspond and work toward the same definitive action through a reciprocal action. Each of these parts cannot change without the others changing as well; and as a consequence, each of them, taken separately, shows and determines all the others.

The anatomist always has to bear in mind how the living being functioned if he wants to understand the laws of organization.

For Cuvier, this general correlation of organs implied that the parts of each living being had some "impossible combinations" and some "necessary combinations": A cutting tooth, fit for tearing flesh, wouldn't coexist in an animal with a hoofed foot, because it wouldn't serve the animal well if it couldn't grasp. There are also necessary combinations: A lion, who eats only meat, has to have sharp eyes and must be capable of pursing prey, seizing it, clutching it, tearing it to shreds, and eating it.

Its organs have to correspond to these functions; hence, its trenchant teeth go along with its sharp claws.

These are the principles that Steno and later Cuvier put into use that we are still applying today. To figure out the names of the bones of the hind foot of a dinosaur, we can compare them with those of a living reptile or bird. So, comparing the bones of the ankle of our Niger dinosaur with those of the ankle of a living ostrich, it was easy to identify, for example, the astragalus and the calcaneum. Furthermore, examining the teeth in the jaws of our African dinosaur, it was easy to decide that this was not a carnivore. Its teeth were lance-shaped, with a relatively thick edge, and were not at all pointed. Applying Cuvier's principle of correlation, I could explain why, for example, the ends of the fingers bore hooves and not claws. So looking at the different parts of the skeleton, making these anatomical comparisons and deductions, showed us whether we were dealing with a herbivorous or carnivorous dinosaur.

Comparisons and correlations can resolve most of our anatomical problems, but studying living animals is not enough to understand the structure and function of some bones in the skeleton of a dinosaur. When the English paleontologist Gideon Mantell discovered the bones of a dinosaur that he eventually called *Iguanodon*, he provided a pretty good reconstruction of this mysterious animal in his first description, given what was understood at the time. But the presence among the remains of a pointed bone, resembling a spine or a horn, eventually embarrassed him. Mantell deduced from his study of the part that it was undoubtedly a horn, and he placed it on the nose of his dinosaur. Later collections, notably the complete skeletons of *Iguanodon* at Bernissart in Belgium, showed us that this was not a horn at all but a paired bone: It was the thumb, modified into a pointed spur whose function is still debated by paleontologists today. By collecting well-preserved skeletons of dinosaurs, we reach a deeper understanding of the anatomical fine points, little by little. But how was Mantell to know at the time?

It must also be said that the principle of correlations cannot always be applied so ideally. Nature sometimes holds some surprises in store for naturalists, and some fossil animals have presented us with a combination of characters that hardly conform to Cuvier's predictions. For example, a fossil mammal discovered at the beginning of the nineteenth century, called a Chalicothere, had teeth much like those of perissodactyls (the hoofed mammals that include horses, rhinos, and tapirs). But contrary to what we might expect, its feet did not bear an odd number of toes, with the middle one enlarged, as in typical perissodactyls. Instead, its toes bore great claws much like those of sloths. Before well-preserved and relatively complete skeletons of these animals were found, it was thought that the teeth and feet belonged to different kinds of mammal.

Furthermore, opponents of the illustrious anatomist pointed out that it would be very difficult to know the bizarreries and complexities of the stomach of a dromedary just by examining its teeth, in fact, even the study of the diverse representatives of living bears could prove perplexing, since all bear species have the same type of teeth, yet some are carnivores whereas others are exclusively fruit eaters. In this case, it really is difficult to deduce dietary regimes from the hard parts alone. But the paleontologist, working mostly on skeletal remains, has no choice, and the approaches and the methods of study that Cuvier set forth remain irreplaceable tools for piercing the secrets of extinct worlds.

At the end of this long and intensive comparative work, I could identify each bone of our dinosaur. The quality of preservation of each piece and the fact that the skeleton was so complete made my task much easier. Information from the excavation and our detailed field records also furnished precious bits of help. A new phase of work now began: I could compare each bone of this dinosaur with the homologous bones of dinosaurs that had already been discovered. I could judge how original my discovery was, and I could know what its relationships were with other, known dinosaurs.

With other dinosaurs? But which dinosaurs? How and why can a species of dinosaur be defined? Why are there genera and species? How does one go about naming a dinosaur?

Since the eighteenth century, naturalists have used the method advanced by the Swedish botanist Carl von Linné, or Linnaeus as his Latinized name was most widely known. His method allows us to use a universal language for all the millions of plants and animals that populate the world. For example, the dog is a *chien* in French, a *Hund* to the Germans, a *perro* to the Spanish, and an *ahkjoun* to the Berbers. Before Linnaeus it was very difficult, even impossible, to describe the diversity of life in a language understood by all people, whatever their own languages were. The genius of Linnaeus was that, by using Latin, the scientific language of his times, he was able to create a system that allowed everyone to classify organisms in the same way. In Linnaeus's time, the world was seen as fixed: It was only a matter of compiling the list of well-defined types that had resulted from divine creation. In his *Systema*

Plantarum of 1753, Linnaeus drew up a list of 5,250 plant species, and in his *Systema Naturae* of 1758, he distinguished 4,235 species of animals. Cuvier, who also thought that species were fixed and did not change, later gave a definition of the term *species* that conformed to this philosophy: "A species is the collection of all organized beings descended from each other or from common parents, and of those that resemble them as much as they resemble each other." Of course, Latin had been the language of scholarly communication for centuries, but with Linnaeus's system, all the individuals of the same species would be dubbed with a unique Latin name composed of only two terms – not a long phrase describing all the organism's attributes. So, the dog was called *Canis familiaris*, the cat *Felis cattus*, and so on. The great contribution of Linnaeus was to provide the first great inventory of living species known in the eighteenth century, and his cleverness was to follow this task of the taxonomist in setting out the *genus* in the same way. The genus (plural, *genera*) comprises species that were regarded as very close to each other, based on their resemblances. So the dog, the wolf, and the fox, which are three different species respectively called *Canis familiarus, Canis lupus,* and *Canis vulpes,* are part of the genus *Canis.* Linnaeus did not limit himself to classifying species and genera; he grouped the genera into orders that were themselves grouped into classes, which in turn were grouped into a kingdom. This was the highest category in Linnaeus's classification; he distinguished the animal, vegetable, and mineral kingdoms.

Today, we have abandoned the concept of the fixed species, and the acceptance of the evolution of species through geologic time has led naturalists to adopt a *biological species concept,* constituted of populations of individuals that form a genetic unity: "Species are groups of natural populations capable of interbreeding and that are reproductively isolated from other similar groups." Paleontologists know better than anyone the reality of evolution, thanks to their field collections and researches, which measure the progression of evolutionary history in the archives of the Earth. But paleontologists unfortunately cannot appreciate which among their fossils were the "groups of natural populations capable of interbreeding." Dinosaurs have been dead for a long time, even if the presence of their bones throughout the Mesozoic Era testifies that they were perfectly capable of reproducing at the time. Paleontological species are real species, even though their skeletal remains leave us far less information than if we had access to their living forms. These species are diagnosed by morphological characters, but when we compare fossil species and living species in a group that still has living representatives (the mice or horses, for example), we find that the characters taken only from skeletal remains can still diagnose and distinguish species of extinct animals as well as they can for today's species.

Paleontologists, like any other systematists, define species and genera. The first discovered species of *Iguanodon* was named *Iguanodon mantelli* in honor of its discoverer, Gideon Mantell. The second discovered species was named *Iguanodon bernissartensis* to evoke the site of its discovery, Bernissart, a small community in the coal basin of Belgium. Then another species was described, *Iguanodon atherfieldensis,* from Atherfield in England; then *Iguanodon dawsoni* in honor of Dawson; then *Iguanodon orientalis,* a species discovered in Asia; and so on. All these species share a certain number of common characters and close resemblances; this is why they are grouped into the same genus, *Iguanodon.* The genera that resemble *Iguanodon* the most, *Camptosaurus, Probactrosaurus,* and *Craspedodon,* were grouped with it into the same family, the Iguanodontidae (the *family* is a grouping that taxonomists introduced after Linnaeus's time because the number of genera became much larger than originally foreseen). The Iguanodontidae were grouped with other families of closely related dinosaurs, the Hadrosauridae, Hypsilophodontidae, and Heterodontosauridae, into the same suborder, the Ornithopoda (the dinosaurs with "avian feet"), which was itself grouped with other suborders, including the Stegosauria (plated dinosaurs), Ankylosauria (armored dinosaurs), and so on, through the order Ornithischia ("bird-hipped" dinosaurs).

In this first stage of comparative work, I needed to show whether this African dinosaur, which was clearly similar to various species of *Iguanodon* that had already been discovered and named, belonged to a new species or even to a new genus. Another phase of inquiry was about to begin. But before getting to that, several prior questions had to be resolved. For instance, how could we establish that the skeleton discovered at Gadoufaoua really belonged to a dinosaur? The great dimensions of the bones were not sufficient evidence: Mammoths and whales have attained impressive dimensions, and their fossil skeletons are also found. In reality, the examination of the various pieces in the field allowed us to come very quickly to the following initial conclusions. The nearly uniform teeth (no separation into incisors, canines, and molars), implanted in the jaws in sockets or *alveoli* (without which there would have been cement between the teeth and the jaws), are characteristic of the

reptiles called Archosauria, which include crocodiles, dinosaurs, and flying reptiles. Examination of the forelimb of the skeleton collected at Gadoufaoua, a limb that was not transformed into a wing, allowed us to
eliminate the flying reptiles nicely. Examination of the limb bones, such
as the femur and humerus, showed us that the shafts were straight, so
we couldn't confuse the bones of our reptile with those of a crocodile.
Finally, study of the arrangement of the bones of the pelvis (ilium, ischium, and pubis) showed that these bones formed a socket at their junction that received the head of the femur. This socket was largely open
to the internal side, which confirmed that the animal belonged to the
dinosaurs, and not to one of the many groups that lived before them
and alongside them in the Age of Dinosaurs. This difference in the hip
socket correlates with the differences in posture. In primitive reptiles,
including the ancestors of dinosaurs, the limb is partly erect: The femur
is oblique compared to the body of the animal. It can be nearly horizontal in lizards and other animals that crawl, which explains the name *reptiles* ("reptant" means "creeping"), given to all animals that have scale-
covered skin and creep on the ground. In dinosaurs, however, the femur
is held straight, parallel to the body as it is in birds, as well as in humans
and other terrestrial mammals.

A closer examination of the bones that form the hip girdle tells us to
which group of dinosaurs our specimen belongs. Dinosaurs are effective-
ly classified into two distinct groups according to the architecture of their
hips: those whose pelvis is *triradiate* (three-pronged), the Saurischians,
and those whose pelvis is *tetraradiate* (four-pronged), the Ornithischians.
In the first group the ilium is above, the pubis is in front, and the ischium is behind, as it is in crocodiles – from which the name *saurischian,*
signifying "saurian pelvis," comes. ("Sauria" is an old name that includes
all the reptiles except turtles.) In the second group, the pubis is more
complicated and has not only an anterior branch but also a posterior
branch parallel to the ischium. It turns out that the posterior or back-
turned branch is the original one, which has simply rotated backward
to lie against the ischium. The anterior prong, which goes up and outward instead of down and inward like the original pubic prong did, is a
secondary development. The posterior pubic branch lying alongside the
ischium is a little reminiscent of the situation in birds; hence the name
ornithischian or "bird pelvis" given to this second group. The terms Saurischia and Ornithischia were chosen by the British paleontologist See-
ley in 1887; these two great subdivisions (i.e., orders) group all the fam-

ilies of dinosaurs. Among the Saurischians are the carnivorous dinosaurs and all the great quadrupedal herbivorous dinosaurs, the sauropods; the Ornithischians include the bipedal herbivores called ornithopods, the horned dinosaurs, the dome-headed dinosaurs, the plated dinosaurs, and the armored dinosaurs.

The pubes of the animal that we collected in the Ténéré were particularly well preserved, and they still articulated with the other bones of the pelvis. The forward branch of the pubis formed a large vertical palette, as thin as a piece of cardboard. The backward-pointing branch lay tightly alongside the ischium. This was unquestionably an ornithischian pelvis. But even if I had not had such magnificently preserved parts, another region of the skeleton could have equally furnished the decisive information. Among the skull bones we found a small, unpaired, horseshoe-shaped bone, the predentary, which fits on the front of the mandibles (or lower jaws) to connect the two front ends of the dentaries. This predentary bone is characteristic of all the Ornithischians.

Our dinosaur, then, was clearly an ornithischian; but this vast group includes more than a dozen families. I would have to find other characters that would help me to understand its relationships with one or another of these subgroups. Because its teeth had thick enamel on only one side, it could not be a plated dinosaur (stegosaur) or an armored dinosaur (ankylosaur); if this had been the case, we probably would have found some evidence of plates or armor at the site. The skull of our specimen was neither thickened nor provided with horns, so I could respectively eliminate pachycephalosaurs and ceratopsians, concentrating my efforts on the four remaining groups: Heterodontosauridae, Hypsilophodontidae, Iguanodontidae, and Hadrosauridae. These are the four main groups of the Ornithopoda. The heterodontosaurids, as their name suggests, have slight differences among their teeth; one tooth is fanglike, which is unusual in dinosaurs. Hypsilophodontids are usually smaller than the specimen that we found, and they still had some teeth in the front of the jaw; and hadrosaurs, or duckbills, have great dental batteries made up of hundreds of teeth with very characteristic, lozenge-shaped enamel surfaces. There remained the Iguanodontidae, and as it turned out, the teeth of the dinosaur from Niger almost exactly resembled the teeth of *Iguanodon.*

A great number of skeletal characters of the dinosaur from Niger could be found among the representatives of the family Iguanodontidae. I made a list of them that included the following: the premaxillaries had no teeth, and the cheek teeth were aligned into a single row; there were 11 cervical (neck) vertebrae, 17 dorsal (back) vertebrae, 6 sacral (hip) vertebrae; the upper edge of the ilium was slightly curved toward the outside, but there was no antitrochanter (the marked prominence to which the muscles that raise the leg attach); the humerus was 55 percent as long as the femur; the tibia was 90 percent as long as the femur; a pointed, spurlike thumb was present; digits II and IV of the hand had three phalanges.

I plunged feverishly into all the publications that dealt with the discovery and anatomy of iguanodontids. I reviewed the characters of all the genera that had already been described – the camptosaurs, *Iguanodon, Craspedodon,* and so on – and it was relatively easy to arrive at a preliminary conclusion. My Niger dinosaur was closest to the genus *Iguanodon;* they shared the greatest number of common characters. However, there was a whole series of differences that prevented the two from being considered identical.

At this point, we must move our story to Belgium. As it happens, one of the world's great venues for dinosaur paleontology is found in Brussels, within the walls of the Royal Institute of Natural Sciences of Belgium. In this magnificent museum is a herd of no fewer than 31 *Iguanodon.* This herd includes 12 more or less complete skeletons and eight portions of skeletons exhibited as they were found in the quarry, as well as 11 skeletons mounted on a framework of metal and restored in the posture that the animals used when they ambled around the Early Cretaceous countryside of Belgium – at least, in an attitude thought to be correct when the animals were first studied and mounted.

I took the train from the Gare du Nord in Paris to bring me to this sanctuary, or rather funerarium, of dinosaurs. The spectacle that awaits the visitor to this museum is absolutely exceptional. The *Iguanodon* are collected in an immense, magnificent room dating from the beginning of the nineteenth century. Two great glassed cages maintain the skeletons in an environment sheltered from harsh variations in temperature and moisture. The bones have pyrite's disease, an impregnation of iron sulfide (FeS), and they would be completely oxidized and rapidly reduced to powder if they weren't kept in their anhydrous (water-free) atmosphere. To avoid their destruction, the bones have also been impregnated with a mixture of alcohol and rubber lacquer. Enclosed within their glass walls, the *Iguanodon* of the Brussels museum seem to be caged, as if they might escape and trample the visitors.

The abundant comparative material collected in this room would allow me to synthesize all the resemblances and differences between the African iguanodontid and its European cousin. Thanks to the understanding and help of the officials of the museum, I received permission to enter one of the great glass boxes that held the mounted skeletons, and I really had the sensation of going in like a lion tamer. But it was with a great deal of respect and humility that I approached these somber bones. This collection, which is unique in the world, has a real history, and it is worthwhile to recount some of the circumstances that accompanied the discovery of these dinosaurs and their presentation to the public.

It all began at the end of the month of March 1878, when Jules Créteur, a miner who worked for the Hainaut coal company at Bernissart, discovered at a depth of 322 meters some bones that were full of pyrite, a pale yellow mineral often confused with gold (it is called "fool's gold" in English). The gallery of the "Luronne" mine crossed a clay pocket that was formed by the collapse of a natural shaft, or *cran* ("notch") in local parlance, made of overlying beds, which hindered the collection of the coal seams. Faced with this unforeseen circumstance, the chief engineer of the coal company wrote on April 12, 1878, to Monsieur Dupont, director of the Museum of Natural History in Brussels, the following telegram:

Important discovery bones in mine Bernissart Coals – decomposing from pyrite – Send de Pauw tomorrow to arrive at Mons station 8 A.M. – I'll be there – Urgent. Gustave Arnaut.

The following day, Monsieur de Pauw, the museum's preparator, descended to the bottom of the mine in Bernissart and discovered a dinosaur foot, which he brought back to the surface. The bones were very fragile and friable, but the handy technician knew how to preserve the parts by impregnating them in glue. Some important excavations then began with the permission of the coal company. For many months, Monsieur de Pauw, a second preparator, a museum guard, and nine miners – despite a thousand difficulties that included a cave-in and a flood – succeeded in extracting a great quantity of blocks full of fossil bones. These blocks, 50 cm to two meters long ($1\frac{2}{3}$–$6\frac{1}{2}$ ft.), were jacketed in plaster before being brought to the surface. After three years of work, six hundred blocks weighing a total of 130 tons were taken to Brussels. All this material was disengaged bit by bit from its matrix, and today the reconstructed skeletons are definitively installed in the museum. A young naturalist from the Brussels museum, Boulenger, showed that this herd of *Iguanodon* was more robust and stockier than the celebrated *Iguanodon mantelli* from Great Britain, and he proposed the name *Iguanodon bernissartensis* for the new species. The detailed study of all this material was undertaken by the master hand of Louis Dollo, a Belgian of French origin and a mining engineer who was named conservator of the Royal Museum in 1882. Among other remarkable studies, Dollo showed that the herd of *Iguanodon* from Bernissart comprised one individual of *Iguanodon mantelli* and 30 individuals of *Iguanodon bernissartensis.*

The discovery of this extraordinary deposit of dinosaurs, its very unusual location more than three hundred meters below the surface, the number of skeletons, and their exceptional preservation caused a great stir throughout Europe at the end of the last century. The removal from Bernissart to Brussels of the small mining town's unusual patrimony was very much resented locally, and a charming song in the Walloon dialect, "The Iguanodon's Complaint" (original words and music by Marcel Lefevre, written in 1912), mentions these striking facts. For the natives, the transfer to the capital of these regional treasures coincided with the decline of coal mining in the Hainaut district:

> Have pity on an Iguanodon
> Who was gracious and sweet
> Who had the muzzle of a rhinoceros
> And skin on his bones
> Whom we dug up, as it happened,
> From the ground quite close to Bernissart
> Where he had perhaps bought
> A nice cemetary plot.
>
> Have pity on an Iguanodon
> Whose history lacks precision,
> On account of which a lot of savants
> Have been saying dumb things for thirty years
> Ah! If only Barnum had found him!
> He would have informed us better;
> I'm sure that if he had looked,
> He could have shown us a living one.
>
> Have pity on an Iguanodon
> Who by birth is a Walloon,

Who for three thousand years had more than
Three hundred meters of Borain humus on his back;
It will infuriate the Flemish nationalists
That the Walloon saurians were so large,
Since no one's found anything in Flemish soil
Except skate and herring bones.[1]

The deposits of Bernissart and its herd of *Iguanodon* were the object
of further curious greed. Beginning in 1915, during the First World War,
the Germans tried to find and collect other skeletons. They ordered the
mine officials to dig a new gallery to reach the fossil bed. But the Belgian miners knew so well how to slow down the work and cause break-
downs in the water pumping system that the end of the war came without the extraction of any new specimens of *Iguanodon*.

Armed with a notebook and a tape measure, I spent long hours ob-
serving, taking notes, and measuring in the middle of the herd of *Iguanodon*. *Iguanodon bernissartensis* truly was more robust than the animal
from Niger, which had a higher skull and protruding nasal bones. Its
muzzle was duckbilled, not parrot-beaked; its predentary was large and
flattened; the front part of the dentary had no teeth, and the neural
spines of the dorsal vertebrae, that is, the apophyses that rise vertically
above the body of each vertebra, were extraordinarily elongated and
gave the vertebral column a very characteristic and special appearance.
In fact there were many other differences in the bones of the pelvis, the
femur, and so on. The additional presence of a specimen of *Iguanodon
mantelli* in the same museum further permitted me to establish that
there were many other differences between this latter specimen and the
one that I was describing. In fact, in pursuing my research and comparisons, I found with a joy mixed with pride that my first dinosaur, though
resembling the *Iguanodon* species of Europe, was unknown up to now,
that its characteristics made it an animal new to science – a new species
of dinosaur and probably even a new genus of dinosaur, so substantial
were the differences between this African fossil and its European relatives.

---

[1] Bernissart is a small town in the Walloon region of Belgium; the small province of
Borinage is also found there, from which "Borain" humus would come. The French-
speaking Walloons have a long rivalry with the Dutch-influenced Flemish region of
Belgium; hence the pointed barbs.

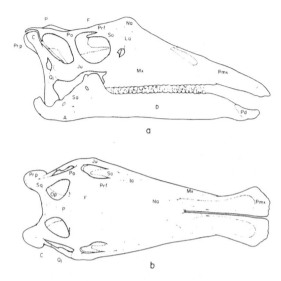

Figure 6. The animal from Niger, *Ouranosaurus nigeriensis:* reconstruction of the skull in lateral and dorsal view. (Drawing by F. Pilard)

I returned to Paris with joy and excitement. Now I had to find the answers to the following questions: What are the closest relatives to this new dinosaur? Which were its probable ancestors? Who were its descendants?

The world of the iguanodontids essentially dates to the Early Cretaceous; most of the species that belong to this family were collected from beds 100–130 million years old. The iguanodontid from Niger was no exception: It was found in a bed of sandy clays of Aptian age, estimated at 110 million years old. The Aptian is what we call a geologic *stage.* All the outcrops of the same age are assembled in this category. Geologists have chosen a *type locality* for each stage, where the geologic beds are well exposed and the deposits are highly representative of the time in question. In this case, they chose some very visible beds quite close to the little city of Apt in the Vaucluse region of France, and so the Aptian is the stage of reference for this interval of time under consideration. This stage name is used by geologists all over the world. The characteristic beds deposited just before those of the Aptian were defined in the same way on the basis of a type locality. This was in the territory of the community of Barrême, also in the southeast of France, where

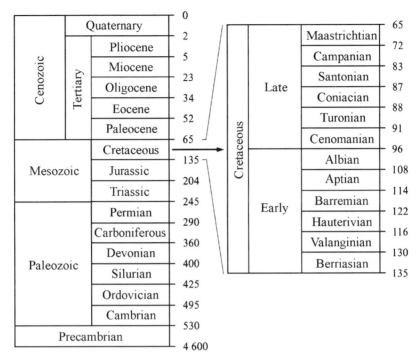

Figure 7. A simplified table of the geologic time scale.

the stage preceding the Aptian, called the Barremian, was defined. The Barremian beds were deposited 115 million years ago. France, Great Britain, and Germany have most of the localities that became the basis for geologic stages, where you can still see the official beds that are called *stratotypes*. This is easily explained, because the origins of geology – and in particular the branch called *stratigraphy,* the study of strata or rock beds – took place in Europe, essentially in these three countries.

The *Iguanodon* of Bernissart were found in a sandy, clayey pocket hollowed out in the more ancient deposits of the Carboniferous period. These sandy and clayey sediments are terrestrial; they were deposited on the continent over a rather long period of time, and so it's fairly difficult to specify their exact age within the series of different beds of the Early Cretaceous. They are probably a little older than those of Niger and their age might be Barremian.

The world of the Early Cretaceous iguanodontids was succeeded by the world of the hadrosaurs, or duckbills, of the Late Cretaceous. Studying all the characteristics of this African iguanodontid, and drawing up a list of the particular characters of hadrosaurs, I was struck by a number of resemblances in the detailed anatomy. Just as in the hadrosaurs, my animal's skull was raised in the frontotemporal region (basically, the forehead and temples), with a convex skull roof level with the nasals. This gives the skull a very characteristic profile that isn't found in the other iguanodontids. Moreover, the highly raised, well-developed, palette-shaped fork at the front end of the pubis, the straight shaft of the ischium (which is expanded at its end into a "foot"), the presence of a protuberance on the upper border of the ilium, seeming to hint at what is called the *antitrochanter* (which is well developed in hadrosaurs and serves as an area of attachment for the muscles that raise the leg), the form of the snout, with its elongated and enlarged premaxillaries – all this strongly evoked the duckbilled dinosaurs. Everything seemed to indicate that this new African dinosaur, which on the one hand resembled the *Iguanodon* species of Europe in a great many characters, particularly the anatomy of its teeth, also showed many features that approached hadrosaurs. To be clear in my own mind, I had to examine some representatives of these hadrosaurs more closely. The problem was that all the complete, well-studied, and accessible specimens of hadrosaurs were found in North America – in Canada and the United States.

There was only one thing to do. In 1971, I left on a five-week trip to meet North American dinosaurs. It was fascinating to discover the great American and Canadian museums that were so sumptuous and full of dozens of complete skeletons of dinosaurs, admirably presented to the public. Lively discussions with the great specialists on dinosaurs nourished my thirst for knowledge and understanding. I fulfilled the dream of every paleontologist and dinosaur lover, visiting the American Museum of Natural History in New York, the Yale Peabody Museum, and the museums of Chicago, Philadelphia, Denver, Laramie, Provo, Toronto, and Ottawa. This tour confirmed what I had thought: Not only did my iguanodontid belong to a new genus and species, but it was probably more highly evolved than the European *Iguanodon*. Its characteristics showed the transformations by which the iguanodontids gave rise over the course of their evolution to their descendants, the hadrosaurs. Furthermore, the stratigraphic position, that is, the place in the geologic column situated in the interval of time that the African dinosaur occupied, fit this hypothesis perfectly: This highly derived iguanodontid was more recent than those of old Europe, but it was older than the duckbills of the Late Cretaceous.

One of the privileges of the naturalist is to be able to name the new species that he discovers. Geographers and navigators can name islands, rivers, and mountains, but this is a more reduced task today because most corners of the world have been explored. In contrast, botanists and zoologists can still describe a great number of new species, because the inventory of the diversity of the living world, past and present, is far from finished. The paleontologist shares with them this opportunity because there are still many fossil species to discover. At the end of my study, this African dinosaur, this iguanodontid from Niger, would finally have a name.

To bring a new genus or species name of a dinosaur to the baptismal font requires a respect for precise rules. These rules, whose origins, as we have seen, go back to the first truly scientific works of classification, and particularly to the work of Linnaeus, were codified little by little and today are accepted by the entire scientific community. Furthermore, they're published in a small, bilingual French–English work, *The International Code of Zoological Nomenclature*. The validity of a new species depends on respect for these rules.

The first step in the definition of a new species consists of choosing a characteristic specimen that is called the *type specimen*, and that will ultimately serve as a reference specimen. With the Bernissart *Iguanodon*, for example, Boulenger had to choose a skeleton from among the 30 that was complete and the most representative among the herd that had been discovered. Then you have to assemble the argument for its novelty; an inventory of all the characters of this new form must be prepared. This list constitutes what is called the *diagnosis:* The sum of these characters allows us to differentiate all the individuals of this new species from those of other species already described. If the characters observed in the skeleton of the Niger dinosaur had been sufficiently similar to those of *Iguanodon bernissartensis*, the Niger specimens would have had to bear the name of the Belgian species. There is an absolute respect for priority: A new species will only be valid if it clearly carries evidence of being new. Its diagnosis must be made public, accessible to the scientific community; that is, it must be published in a journal and accompanied by an illustration. The date of publication of the description of the new species serves as the date of reference. Finally, it has to get a name that respects the rules established since Linnaeus, composed of two Greek, Latin, indigenous, or Latinized terms.

Accordingly, I had decided to name the new species *Ouranosaurus nigeriensis*. The first term was taken from the Arab word *ourane,* which signifies "valiant, courageous, bold." This name was given by the Tuaregs of Niger to the sand monitor lizard. For the Tuaregs, this great lizard recalls a legend about their maternal uncle, who was metamorphosed by God to punish him for having killed a camel belonging to Noah. I don't know if this wonderful legend, related by Alibert, a famous camel driver, is true, but the great fossil reptile discovered in the sand-covered regions of the Ténéré, which we dug out in front of several Tuaregs on their way to Gadoufaoua, doubtless evoked for these natives of Agadès some giant *ourane* enclosed in these beds since time immemorial.

The suffix *-saurus* is taken from the Greek word *sauros,* which means saurian or lizard, and the word *nigeriensis* alludes to Niger, the country where it was discovered. The type specimen is the complete skeleton that we collected and that carries the number GDF 300. The work of description, diagnosis, and illustrations that attest to the creation of this new species of iguanodontid were published in 1976 in a memoir published by the CNRS. For the scientific community, this new species of dinosaur will carry from that time forward the name *Ouranosaurus nigeriensis* Taquet 1976. This very simple system allows each species to be known by the date of its creation and the name of its inventor. Thus *Iguanodon mantelli* was first named by Meyer in 1832, and *Iguanodon bernissartensis* was first named by Boulenger in 1881.

On occasion, later research shows that certain genera or species belong to genera or species that have already been described and named. The researcher who brings this evidence to light through his work thus modifies the status of these genera or species. For example, it was shown in 1903 that the name of the celebrated *Brontosaurus,* so familiar to children as well as to adults, had been given to some remains of a large herbivorous, quadrupedal sauropod dinosaur that really belonged to the genus *Apatosaurus.* The kicker is that these two names, *Brontosaurus* ("thunder lizard") and *Apatosaurus* ("deceptive lizard") were both given by the same researcher, the great American dinosaur hunter O. C. Marsh. In 1903, Elmer Riggs, of the Field Museum in Chicago, restudied the two sauropods described by Marsh. The genus *Brontosaurus* had been based essentially on the features of the shoulder girdle and on the presence of five vertebrae in the hip girdle. Riggs showed that these pieces had the same features as those corresponding to the genus *Apatosaurus,* which Marsh had previously described. Only their dimensions were different, because the specimen named *Apatosaurus* was simply a juvenile form of the one that was represented as an adult by the specimen that Marsh called *Brontosaurus. Apatosaurus* had been described and named before *Brontosaurus,* and in view of the rule of priority, *Brontosaurus* became a junior synonym of *Apatosaurus* and had to be abandoned. The problem is not always so simple or so easy to resolve, and it sometimes requires submission of the parties to an international authority that is charged with judging and rendering a decision, at the end of a long inquest and the assembly of a solid dossier with all the contradictory arguments.

Ouranosaurus nigeriensis, the ouranosaur, since that is what we began to call it, is thus a new iguanodontid dinosaur, an African cousin of the *Iguanodon* of Europe. But it is a genus with certain characters that also evoke the descendants of the iguanodontids, the duckbilled dinosaurs or hadrosaurs. The ouranosaur is less robust than *Iguanodon bernissartensis;* it's closer to *Iguanodon mantelli,* which is more gracile, but we've seen that its skull characters and the long neural spines on its vertebrae set it apart. In the Gadoufaoua region, some dozens of meters from the skeleton of *Ouranosaurus,* we found and collected the remains of a second dinosaur whose preparation is now finished. This second skeleton is also of an iguanodontid, but curiously, all the osteological characters of this iguanodontid, which is more massive and stocky than the ouranosaur, are much closer to those of *Iguanodon bernissartensis.* The forward part of the pubis is slightly elevated, the ischium is curved, the ilium has a short process in front of the hip socket, the hand has a first finger reduced to a large conical spur, and the neural spines of the vertebrae are short. Though the resemblances between this robust iguanodontid and *Iguanodon bernissartensis* are strong, the form from Niger is not identical to that from Europe because it has a forelimb that is very robust, with a short but very thick forearm (radius and ulna); also, the wrist is very large and the hand bones are short and large, seeming to indicate an animal that was permanently quadrupedal.

This double discovery is astonishing: There were two contemporary iguanodontids in Niger! The gracile one is relatively close to *Iguanodon mantelli* and the robust one is closer to *Iguanodon bernissartensis.* In Europe, the differences between *I. bernissartensis* and *I. mantelli* are not extremely marked, so many speculations have been raised on the subject of the two species. Is one adult (*I. bernissartensis*) and the other juvenile (*I. mantelli*), or is one male and the other female? In the Gadoufaoua beds, the differences between the gracile form (*Ouranosaurus nigeriensis*) and the robust form are such that it is completely impossible to confuse the bones of one with those of the other, and it seems implausible that one is the male or the juvenile and the other is the female or the adult.

The discoveries in Niger thus tend to support the hypothesis of the existence of two related groups of iguanodontids that lived in the same Early Cretaceous communities in Europe and in Africa, one comprising gracile forms and the other robust ones. At the beginning of the Early Cretaceous, the robust form (*I. bernissartensis*) is very close to the gracile form (*I. mantelli*). At the end of the Early Cretaceous the differences between the robust and gracile forms are more important. If my hypothesis is correct, the presence of two groups of iguanodontids in the Early Cretaceous justifies *a posteriori* the separation that Louis Dollo made in 1882 of the *Iguanodon* specimens from Bernissart into two distinct species: the massive form and the gracile form. My hypothesis logically excludes the idea that some of these forms are adults or males and others are females or juveniles. If there is still some question about the forms from the beginning of the Early Cretaceous, there can no longer be any doubt by the end of the Early Cretaceous, when the differences between the gracile and robust forms have only been accentuated over time. New detailed studies of the Belgian *Iguanodon* specimens by the British paleontologist David Norman have strongly confirmed the presence of two distinct species of *Iguanodon* at Bernissart.

I presented these ideas about the evolution of iguanodontids and the origin of the hadrosaurs in 1973 to my French and foreign colleagues at an international colloquium of vertebrate paleontology that was held in Paris, organized by the paleontology laboratory of the Museum with the support of the CNRS. In my communication, I tried to understand the reasons for the replacement of the iguanodontids by the hadrosaurs. The hadrosaurs were characterized by jaws provided with many parallel rows of teeth, veritable crushing dental batteries that were very different from the jaws with only two or three tooth rows found in iguanodontids. This substantial addition of teeth (there are many hundreds in the mouth of a hadrosaur), correlated with the presence of dental plates, implied a change in the dietary habits of this lineage of dinosaurs. The iguanodontids would have eaten leaves from cycads, ginkgos, and araucarias. Why didn't their descendants do the same? My hypothesis was that these jaw transformations coincided with the appearance in the course of the Early Cretaceous of the first flowering plants. The first pollen grains from flowering plants are known from the Barremian and Aptian deposits. Among these flowering plants are found the first grass-like plants, and we can infer that the vegetation of the Early Cretaceous underwent a slow but profound transformation. The great forests of araucarias must have ceded their place, a little at a time, to more open and herbaceous plant associations. The iguanodontids and their descendants had to adapt to these new conditions of life. So the evolution of these herbivorous ornithopod dinosaurs was tied to that of the Cretaceous flora. Large dental surfaces adapted to crushing and grinding vegetable matter replaced the fine cutting jaws of their iguanodontid ancestors; the success and the expansion of the Hadrosauridae were probably due to these adaptations to a new dietary regime.

What happened among the ornithopod dinosaurs of the Cretaceous happened again during the course of the Cenozoic Era (the Tertiary period) with the herbivorous mammals that were the ancestors of horses. Living horses descended from remote ancestors that ate leaves. With the expansion of grasslands their teeth were modified, and the evolution of enamel crests on their molars allowed them to consume the tough herbaceous vegetation of the steppes and savannas with maximum efficiency. The transformation of their limbs enabled them to run rapidly, at a gallop. Hadrosaurs, for their part, remained committed to more humid climates, and their limbs never transformed like those of the horses. But their jaws were certainly well adapted for food processing.

Since my description of the ouranosaur, which seemed an iguanodontid by its teeth but a hadrosaur by its snout and nasal arches, many works on all the representatives of these bipedal ornithopod dinosaurs have been published by my English and American colleagues, David Norman, David Weishampel, and Jack Horner. In their works, the ouranosaur finds itself right in the middle of their reflections on the evolution of iguanodontids and the origin of hadrosaurs.

Research on the possible relationships among living organisms is the object of a scientific discipline that is practiced by all naturalists: *systematics*. As my colleagues at the Museum, Loïc Matile, Pascal Tassy, and Daniel Goujet, noted in 1987, systematics is the study and the description of the diversity of living things, the research into the nature and causes of their differences and resemblances, the analysis of their relationships to each other, and the development of a classification that reflects these evolutionary relationships. Since Linnaeus, systematists have used several major criteria to group organisms and to try to order these groups: general resemblance, ecological resemblance, and, when possible, characters that seem obvious, such as the existence of feathers in birds. When the theory of the evolution of species was developed, first by Lamarck and then by Darwin, one of the tasks of the naturalists was to seek out the genealogical relationships among living species, which were the fruit of this long historical process of evolution through the course of geologic time.

Systematics thus became a discipline that took into account the modification of characteristics in the course of descent. Then a great revolution in methodology took place among systematics, thanks to the work of Willi Hennig, a German entomologist who specialized in flies, and who in 1950 published in German his principles of a theory of *phylogenetic systematics*. Instead of looking only for relationships between ancestors and descendants, Hennig proposed to try to understand which among several taxa (several genera or species, for example), were the two most closely related. With this new method, it became possible to study the genealogy of living organisms more precisely, and to draw up trees whose branches would describe the history of the appearance of these lineages. Each living being has a mélange of traits; some traits are primitive and some are specialized, or *derived*. The Hennigian method is devoted to constructing trees that are called *cladograms*, using *only* those specialized, or derived, characters that alone can indicate genealogical history, namely by showing the new evolutionary stages of the organisms under study. *Derived characters* are new evolutionary features shared only by the descendants of the common ancestor that first evolved them; for Hennig and his disciples, these are the only characteristics that can be used in establishing phylogenies, or trees of life.

With the spread of Hennig's work and methods, paleontologists began as early as 1984 to try to find evolutionary relationships among dinosaurs, and particularly among the different species of iguanodontids and hadrosaurids. At the international Symposium on Mesozoic Terrestrial Ecosystems held in Tübingen, Germany, two communications dealt with this subject. One was by Angela Milner and David Norman, the other by Paul Sereno. For Sereno, cladistic analysis showed that among the iguanodontids, and even among ornithopod dinosaurs, *Ouranosaurus* from Niger was the closest relative of hadrosaurids. The duckbilled beak characteristic of hadrosaurs was present in all its essential features

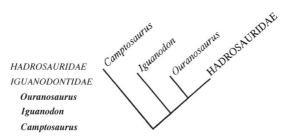

Figure 8. Cladogram showing the relationships among *Ouranosaurus* and some other derived ornithopod dinosaurs, with the traditional classification at left. (After Sereno 1990)

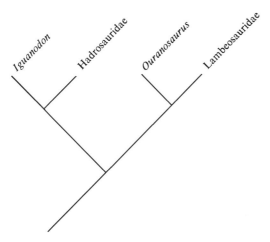

Figure 9. Cladogram showing the relationships among *Ouranosaurus, Iguanodon,* and the two lineages of hadrosaurs, the Lambeosaurids and the Hadrosaurids. (After Horner 1992)

in *Ouranosaurus,* including the nasal depressions and the anteriorly enlarged premaxillary. In a long article in the magazine *Research and Exploration,* published by the National Geographic Society, Sereno identified *Ouranosaurus* as the *sister-group,* or closest known taxon, to all the hadrosaurs, and he placed all of them together in a group that he called Hadrosauroidea.

In 1990 and 1992, my colleague and friend Jack Horner, of the Museum of the Rockies in Montana, published two articles in which he also presented a cladistic analysis with his interpretation of the evolution of iguanodontids and hadrosaurids. His conclusions were considerably different from Sereno's. For Horner, the duckbilled dinosaurs comprised two groups of different origin. They did not form a *monophyletic group,* that is, a group with a single common ancestor, but rather two distinct groups: the hadrosaurines and lambeosaurines. The first was closer to *Iguanodon,* with which it shared a common ancestor; the second was closer to *Ouranosaurus* for the same reason. In this case, *Ouranosaurus* would only be close to one half of the duckbilled dinosaurs.

Still more recently, my colleagues David Weishampel, David Norman, and Dan Grigorescu advanced a detailed description of *Telmatosaurus transsylvanicus,* the only known European hadrosaur, whose presence in France I had shown in 1973. They erected a new cladogram, the result of cladistic analyses, that explain hypotheses about the relationships among several taxa that evolved from the same ancestor (species, genera, families, and so on), which are called *clades.* From their analysis, which dates from 1993, my colleagues concluded to the contrary that *Iguanodon* and *Ouranosaurus* were separate from the lineage that led through the Cretaceous to the duckbilled hadrosaurs. In this case, the specializations seen in the skeleton of *Ouranosaurus* would be only *convergences,* that is to say, resemblances that appear independently in *Ouranosaurus* on the one hand and in the Hadrosaurs on the other. These mistaken resemblances could not be evidence of relationship.

These different examples show that the history of relationships between iguanodontids and hadrosaurids are still the subject of numerous discussions and divergent interpretations, even though different workers are using the same methods (Hennig's). The *Ouranosaurus* is implicated and discussed in all the hypotheses presented here, and its characters are analyzed, weighed, and appreciated. Whether *Ouranosaurus* is the direct ancestor of hadrosaurs or whether its similarities are the result of parallel evolution with hadrosaurs, without having given rise to any descendants, it is still an ornithopod dinosaur that is frequently cited and used by researchers in their studies and comparisons.

In 1973, I published the first reconstructions of the *Ouranosaurus* "on the bone" and then "in the flesh" in a scene from the Early Cretaceous, with the cooperation and talents of Françoise Pilard, the artist from the paleontology laboratory of the Museum. Since then, this new African dinosaur has gradually gained admission to the pantheon of other celebrities such as *Iguanodon* and *Stegosaurus.* The type specimen was placed in the National Museum of Niger in Niamey, inaugurated by the president of the National Assembly of that country, Boubou Hama. A small Niger girl, very timid and cute, with her plaited braids, dressed like an

Figure 10. Reconstruction of an Early Cretaceous scene, showing *Ouranosaurus nigeriensis* with its large dorsal crest, and in the background, a carnivorous dinosaur. (After B. Halstead-Tarlo)

ouranosaur in silk colored like the Niger flag, presented the president with a pair of scissors to cut the symbolic ribbon across the entry door to the brand new pavilion of paleontology. A stamp was issued to celebrate the event. Today the *Ouranosaurus* is illustrated and cited in all the treatises of paleontology, and a vast number of different reconstructions has been created by a horde of biological artists. Each one carries its own personal touch, but the *Ouranosaurus,* with its imposing crest along its backbone, always remains recognizable in the middle of the immense troop of dinosaurs that we know today. And the father of the *Ouranosaurus,* which these collections and studies have made me, is obviously very proud to see his son, to whom he has devoted so much care, cutting such a fair figure in all the group photos of dinosaurs that have been published in both French and foreign works.

Figure 11. The trackway of a sauropod dinosaur, a huge herbivorous quadruped, on the slopes of Mount Arli, 140 million years old, near Agadès, Niger. (Photo by P. Taquet)

CHAPTER THREE

■

TRACKING THE DINOSAURS

THE FOSSIL RICHES of Niger are not limited to the dinosaur skeletons of Gadoufaoua. On March 20, 1966, the prospecting geologists of the Atomic Energy Commission took a crew of the Museum's paleontologists, including Léonard Ginsburg, Bernard Loiret, and myself, on a little geological excursion. We were joined by Albert-Félix de Lapparent, the pioneer of dinosaur research on the Saharan platform.

Our little crew boarded two Land Rovers and left the town of Agadès early, taking the trail to the west that led to In Gall and Tahoua. Across nearly 150 kilometers stretched an immense, perfectly horizontal plain, the Irhazer, made of clay beds that were laid down at the end of the Jurassic and the start of the Cretaceous periods. This plain is covered by a fine layer of recent clay, because it becomes a floodplain once a year, during the rainy season. This is when the *wadis* (gully oases), swollen with moisture blown in from the nearby Aïr mountains, deposit their sediments. In July, this gray plain, blackened, dessicated, and powdery, is transformed as if by a miracle into a vast, verdant, and welcoming prairie. Each year, rushing in from all directions, come an assembly of nomads, Tuaregs and Peuls, with their immense herds of dromedaries, bororo cattle, sheep, and goats. They come for the "salt diet," the forage abundant and rich in mineral salts, that refreshes the livestock after the long dry season. But this was March, and all we saw were clouds of dust rising from our trail. Everything was dry and broiling; far away, thanks to the effects of great mirages, the heat waves transformed the countryside into a vast stretch of chilly water strewn with fortresses whose towers and scalloped ramparts stood out against the horizon.

■

After rolling along quickly for 60 km (about 40 miles) on this perfectly flat track, we veered southward toward Mount Arli. The word "mount" is a little grandiose to apply to a mound of earth about 20 meters high (65 feet). This mound was formed by the disturbance of the geologic beds when the mountains of the Aïr range to the east were uplifted, bringing the underlying beds to the surface. The beds are made of silts, very fine particles of detrital sediments, forming solid slabs colored blue and black by manganese oxides. These hardened benches alternate with beds of fine, very friable claystone. So the west flank, broken by a tumble of parallel north–south faults, forms a kind of giant stairway, but the solidity of these stacks of rock is somewhat relative. It's like a giant staircase whose steps are losing their parquet layer.

Our geological colleagues dragged us to the second of these steps, apparently just to watch our reactions. They didn't have to wait long. There, in front of us, at our feet, even under our feet, were footprints – enormous footprints nearly 40 cm long (16 in.), perfectly imprinted in the rock. And over there was another, and another, and still another . . . ten, then twenty, then forty prints, lined up in two distinct rows. One trackway was made of the prints of elephantine fore and hind feet that testified to the passage of an enormous animal shortly after the sediment was laid down. Another track had such good preservation and freshness that we were brought up suddenly, and caught ourselves looking around for the author of these perfect footprints. We were amazed and impressed, and as we followed these traces, we had the definite feeling of moving back in time to the world of the Late Jurassic.

A dinosaur passed by here 140 million years ago. Yet its traces were so clean that we could distinguish perfectly the impression of the palm of its hind foot, the grooves left by its four great claws, the trickle of mud that surrounded the track after the foot was impressed, the force of the animal's weight on the ground, and even the texture of the skin of its feet. How could such fossilization be so good? How could these tracks be preserved so well through geologic time?

The answer is simple; the process is unfolding today before our eyes, all over the world. All we have to do is look along the edges of great lakes and rivers in Africa to understand what happened at the exact moment when these dinosaurs were wandering the plains of the Irhazer in Niger. During my third African expedition in 1967, with my colleagues Armand de Ricqlès and Raymond Desparmets, I had the opportunity to observe a truly spectacular example of this process. Passing through the capital of Chad, at Fort Lamy (now called N'Djamena), we reached the edge of the Chari, the beautiful river that flows into Lake Chad, where a herd of hippopotamus was cavorting. On the banks of its great meanders, the river had flooded and deposited a thick bed of yellowish mud before returning to its usual level. Over a vast surface, hundreds of animals, both birds and mammals, had left their footprints perfectly impressed in the mud. A hippopotamus had emerged from the river to take a little walk, and its peregrinations could be read very clearly on the ground. Its fore and hind feet, left and right, were clearly marked, and we could count the number of its toes and the size of its claws. After its little promenade, Chad's persistent sun had dried the muddy surface, hardening and "cooking" the clay to make it a finely textured but very solid bed. A bit later, a little sand and powder, carried by the wind, were deposited on the surface. The track of this hippopotamus would remain just like this until the rainy season, when the Chari, overflowing its banks, would come to cover the traces and protect them from erosion with another coat of mud several centimeters thick. A new muddy surface would then be ready to record the passage of another generation of animals. This alternation of wet and dry seasons shows us the principles of how tracks are preserved, as well as how other traces of animal activity can be recorded.

If these deposits of fine clay are not worn away later by the river's erosion, and if this regular sedimentation carries on for years, great accumulations of these geologic layers, rich in trackways, will be stockpiled in sedimentary basins and preserved for thousands of years . . . millions of years . . . some of them right up to the present. Soon after being laid down – at least, in geological terms, which is to say after some hundreds or thousands of years – the water that percolates through these sediments, charged with mineral salts, will wind up impregnating and soaking into the clay to transform it, through a slow chemical maturation, into a very solid sediment whose particles will be incorporated into a sandy or calcareous cement. Through the course of geologic time this fluid, clayey mud, suitable for preserving footprints faithfully, will become a very solid rock capable of resisting the strongest deformational forces. This collection of processes that affect sedimentary deposits and transform them progressively into sedimentary rock is called *diagenesis*. This is how these forces work ceaselessly at the Earth's surface: deposition, fossilization, sedimentation, compaction, diagenesis, and erosion are the forces that transform and remodel the face of the planet. Their study constitutes an important branch of earth science known as *surface geodynamics*. The fossilization of the hippopotamus tracks today helps us to understand the fossilization of the dinosaur tracks so long ago.

After the euphoria of this initial encounter with these still unidentified traces on the flanks of Mount Arli, our little crew went to work immediately, taking out tape measure, compass, graph paper, and camera equipment. We measured as well as we could the orientation of each beautiful impression, its dimensions, the length and width of each fore and hind footprint, the number, length, and orientation of the grooves that testified to the presence of claws. This magnificent track of a quadrupedal dinosaur took up 60 meters of thirty-one cycles, a cycle being the fore and hind tracks of the same side of the animal. We could distinguish perfectly in each cycle the impression of the foot and its great size (36 cm long, nearly 15 in.), and that of the hand, which was shorter (22 cm long, about 9 in.) and relatively smaller. The last phalanges were directed outward in the fore foot and inward for the hind foot, exactly as in *Diplodocus* (to whom we shall return in a later chapter). The dimensions of this trackway, the spaces between the prints of the left side and those of the right, and the size and the form of each of the prints showed us that we were were dealing with the type of large dinosaur called a sauropod. The sauropods include all the large Saurischian (triradiate pelvis) herbivores, and count among their ranks the *Diplodocus, Brachiosaurus, Camarasaurus, Cetiosaurus,* and many other forms. These are the largest terrestrial vertebrates that ever walked the planet.

It's extremely rare to collect a well-preserved skeleton of these Mesozoic giants, and it's even rarer to find a sauropod specimen with all its foot bones in place. Usually, after the animal dies, the hand and foot bones and fingers and claws all separate from each other. The little round wrist and ankle bones, as well as the last phalanges of the outer digits, which are reduced to small nodules of bone, are difficult to recognize and to collect, so the fine structures of the sauropod hand and foot are still poorly known. This is why the study and interpretation of sauropod footprints, the way their claws met the ground, their phalanges, and the form of their feet are all so precious for knowing and understanding the anatomy of the bones. In the case of this sauropod on Mount Arli, we could see that the first finger of the hand was offset toward the midline of the body and was armed with a huge claw. This observation was unusually valuable, because it was one of the rare cases in which this digit has been found pressed well into the ground. Some of the prints very distinctly showed the traces of five digits on the fore foot. The first three, the inner ones, all had narrow claws, whereas the two outer ones ended in large rounded stumps that clearly bore no claws. The entire impression of the fore foot was large and short, showing that the bones of the palm did not rest on the ground. The animal walked on its toes and not on the flat of its hand.

The track of the foot was longer. The sole bones and the heel were impressed on the ground clearly. Four digits were always distinctly visible. The claw of the first toe was the longest: digits II, III, and IV were also armed with long claws. Digit V, the little toe, usually only rarely seen, was very short, and its claw could be curved outward or even turned backward. The Mount Arli sauropod thus had hind feet in which all the digits bore claws, in contrast to what was known in American dinosaurs such as *Diplodocus* and *Camarasaurus.*

From this trackway we could also estimate the dimensions of the dinosaur, by measuring the distance between the prints of the left foot and the right foot (24 cm, or 10 in.) and the stride (the distance that separates two prints of the same foot), which varied from 1.9 to 2.3 meters (6–7 ft.). But we had to be careful. Actually, the length of a stride won't give us any indication of the sizes of some parts of the dinosaur; some of them, like the neck or the tail, attained considerable length in these large reptiles. The measure that does give a good indication of the proportions of the animal is the distance that separates the shoulder from the hip. (These sockets are called the *glenoid cavity* and the *cotyloid cavity* by anatomists, but we often just call the distance between them the "wheelbase," as for a car.) This distance can be measured easily in quadrupeds: All you have to do is to measure the distance that separates the point midway between two successive prints of the hand from the point midway between two successive prints of the foot of the same side. This method, which is used on living crocodiles, seems valid for sauropods, but of course it's useless for bipedal (two-legged) dinosaurs. This is how we were able to estimate the wheelbase for the Mount Arli sauropod at 2.9 meters (9 ft.). We compared it to the cast of the mounted skeleton of the *Diplodocus* at the National Museum of Natural History in Paris, which is 23 meters long (70 ft.) and has a wheelbase of 3.2 meters (10 ft.). This gave our Niger sauropod a fairly respectable length of 15–18 meters (about 50–60 ft.).

Albert-Félix de Lapparent told us that in February 1954, in the middle of his prospecting the region of In Gall on camelback – about 90 km away from where we were now, in beds of the same age – he had collected some remains of a large sauropod whose spoon-shaped teeth were characteristic of the camarasaurs (as opposed to the pencillike ones typical of the diplodocids, a second group of sauropods whose most famous representative is *Diplodocus*). It was a member of this kind of sauropod, which in 1960 he named *Rebbachisaurus tamesnensis*, that might have made the tracks at Mount Arli. He estimated its length at about 20 meters (65 ft.).

These tracks also showed us that the animal did not drag its tail on the ground. In fact, no trace of this long appendage was visible, although the soft mud preserved even the smallest details of the feet. The absence of any trace of the tail is typical in dinosaurs, and it shows that all these great reptiles, whether quadrupedal or bipedal, walked or pranced with their tails stretched horizontally above the surface of the ground. I know of one rare example of a dinosaur tail mark, described by my colleague and friend, the Portuguese scientist Miguel Telles Antunes. This track was made by an *Iguanodon* who left its imprints on some Early Cretaceous beds south of Lisbon, on Cape Espichel. In this case, the animal was resting, leaning on its hindquarters; one can distinctly see the impression of two hind feet, and behind these along the median axis, a long depression, broader in front than in back, and about 50 cm in length (18 in.). This corresponds nicely to what one would expect for a tail impression. These Portuguese tracks are found at a picturesque site where the slabs that bear them form large cliffs that plunge obliquely into the sea. According to the Portuguese, who are very religious, this place was chosen by the Holy Virgin when she came to evangelize Portugal, perched on the back of a donkey. Popular belief had thus come to attribute these prints to a member of the horse family, but paleontologists had assigned them to a dinosaur. Actually, it was rather nice for us to think that a reptile that had been so often compared to a dragon, that Western incarnation of sin and evil, might have surreptitiously wound up as a substitute for a brave mammal that had carried the mother of Christ.

The Mount Arli track also told us that the sauropod was wandering solo, apart from its companions. At other tracksites, some broad expanses have preserved hundreds of tracks, and these have been inventoried and submitted to statistical calculations, to determine the predominant direction or directions in which the animals were moving. In certain particularly favorable cases, it has been possible to establish the passage of true herds, troops of dozens of individuals, as my colleague John Ostrom of Yale University did for some Connecticut Valley footprints of the Early Jurassic. This led him to advance the hypothesis that these dinosaurs, particularly the prosauropods and sauropods, might have been gregarious, traveling habitually in large herds.

The regular footfalls of the Mount Arli sauropod were even able to tell us that the animal was walking with what we might call a contented gait. We can say this because dinosaur footprints have been used to calculate the speeds of the animals that were making them. A British biomechanician, R. McNeill Alexander, put forward a mathematical formula that can calculate how fast the animal was going. Alexander didn't invent this formula; he found it in the work of the naval architect William Froude. The *Froude number,* oddly enough, is used to calculate the speed of boats. It is equal to $v/(v-gl)$, where v is the velocity of travel, g is the acceleration due to gravity, and l is the length of the limbs. Don't ask me why this formula, which applies to ships, also works for dinosaurs, the vessels of the Mesozoic. I think I can understand that two bow waves in front of ships, or two running animals, or two systems subject to gravity, are similar in dynamic terms when their Froude numbers are equal. Alexander showed that there was a mathematical relationship between the square of the Froude number and the length of the stride, the height of the limb at the level of the hip socket . . . and the speed of a dinosaur. So sometimes we can determine not the total size of the dinosaur but its speed, based on its footprints. Alexander's conclusions are that all the large sauropods ambled along slowly, about as fast as a walking man, around one meter per second. At top speed they would have approached elephants as they walk fast.

So the study of dinosaur tracks brings us considerable information that we couldn't know from the skeletons alone. Thanks to the examination and study of footprints and trackways, we can do some paleobiology: We can reconstruct the appearance, the posture, and gait of these extinct reptiles; we can literally step into their steps and follow them in their peregrinations to learn about their way of life. The study of fossil footprints is the basis of a branch of the discipline of paleontology that is called *paleoichnology.*

Paleoichnology began as a science in 1802 when Pliny Moody, a young boy working with his father on their farm in South Hadley, Massachusetts, brought to light with a stroke of his plow a large slab of rock on which were imprinted five small, tridactyl (three-toed) footprints.

This strange discovery attracted the attention of the Moody family members, and they placed it as a decoration over the entry gate to the farm, where it remained for seven years before being purchased by Dr. Elihu Dwight. For a long time Dr. Dwight showed these imprints to all his visitors, telling them jokingly that these were probably the footprints of Noah's raven made after the Great Flood.

Thirty years later, the slab became the property of Professor Edward Hitchcock of Amherst College. Hitchcock had been interested since 1835 in the collection and study of the footprints found in great numbers in the rocks of the Connecticut Valley. In 1841 Hitchcock described the prints that Pliny Moody had discovered, giving them the name *Ornithoidichnites fulicoides* by virtue of their resemblance to the American coot, a bird that lives today in great numbers in the lakes and streams of North America.

Hitchcock soon became the premier specialist in the study of vertebrate footprints, and in 1836 he published the first inventory of his discoveries. Consistent with what was then understood about the history of Earth, he regarded all the tridactyl prints of the Connecticut Valley, no matter how large or small, as the tracks of antediluvian birds. Hitchcock's "birds" soon became so abundant that a host of slabs bearing superb examples of tridactyl prints ornamented the Appleton Natural History Cabinet at Amherst College.

It was Hitchcock who created the term *ichnology* (from the Greek *ichnos* for footprint or track) for this new branch of paleontology. It was also Hitchcock who applied Linnaeus's elaborate methods of classification to these fossils, creating genera and species according to the characters that he observed in the tracks. In this case, we now speak of *ichnogenera* and *ichnospecies,* and we put them in a classification that is parallel to the one based on skeletal remains. The study of a trackway can tell us that we're dealing with a large quadrupedal dinosaur, such as a sauropod; but it can't tell us the animal's identity – whether we're dealing with a *Diplodocus* or its contemporary *Camarasaurus,* or whether it's a *Diplodocus longus* or a *Diplodocus carnegiei.* Only the presence of the skeleton of the trackmaker can tell us this. In practice, though, we almost never find fossil skeletons in the same beds as the trackways. This is partly why we keep the classification based on footprints separate from the one based on bones, just as there is a separate classification of fossil eggs and eggshells because it is so difficult to attribute a fossil egg to a particular species of dinosaur.

Hitchcock, then, was a great pioneer in the study of fossil footprints, and an excellent detective of the past. But he described a multitude of species of tridactyl tracks, all of which he attributed to birds – incorrectly. In reality, these prints were made by bipedal carnivorous theropod dinosaurs and by herbivorous ornithopod dinosaurs during the Early Jurassic, some 200 million years ago. This was long before the first known appearance of birds in the Late Jurassic, 140 million years ago. However, Hitchcock had a good excuse: His first work was written before the dinosaurs were "invented" by the British paleontologist Richard Owen in 1842. And it was only in 1856 that the first discovery of dinosaur bones was made in North America. By the end of the nineteenth century, Hitchcock's concept of giant birds running around the Early Jurassic plains of Massachusetts and Connecticut had lost all credibility. Since those heroic times, a multitude of dinosaur tracksites has been discovered in a great many countries, and paleoichnology has become such a solid science that its specialists hold international congresses.

Our study of the great trackway of the quadrupedal dinosaur from Mount Arli concluded, we explored the surroundings and soon found a large number of very different prints from those we had just examined, littering the well-exposed slabs. I should say that the low-angled light of the sun, which was then sinking in the west, was a great help: Its oblique rays accentuated the shadows on the ground and brought out every little bit of relief and the most subtle depressions on the rocky surface. Thus all the tracks that had been made in a given area could be discovered much more easily than at midday, when the sun was at its zenith and its direct rays turned the rock slabs into dazzling surfaces that blotted out all relief.

And the discoveries multiplied; over many dozens of square meters there were large tridactyl tracks 35 cm long (14 in.), with a space between them of another 35 cm. These were the tracks of a bipedal carnivore, a theropod, that had a leg length of 1.46 meters (about 5 ft.). These latter tracks were remarkable for the quality of their preservation: We could easily see the scaly impression print of the skin at the level of the claws. Finally, a bank displaced by erosion, located 1.2 meters below the level of our large sauropod track, yielded us about 50 prints of small animals that were curiously slender and spread apart. One of these typical tracks was pentadactyl (five-toed) and measured 5.5×6.8 cm (about $2\frac{1}{4}$ by $2\frac{5}{8}$ in.). The three middle toes were long and lightly curved at their ends; the outer toe was shorter, and the inner toe seemed reduced. This track reminded us, even in its size, of the track left by the foot of a triton or other small salamander, rather than that of a lizard. So it seemed to be more like an amphibian trackway than one of a reptile, but we had no idea about the identity of the track-maker.

After many hours of study, prospecting, discovery, and thinking, the day was nearly over. But before leaving Mount Arli, we set about one last activity designed to bring back an impression – because that's the right word for it – of the magnificent tracks of our sauropod. We had taken the precaution of bringing along some plaster and burlap. This allowed us to make a mold of the best pair of footprints from this spectacular track, one from a fore foot and one from a hind foot. A thin film of soap spread over the footprint on the slab of rock ensured that the plaster wouldn't stick to the surface, and so we were able to take an impression in plaster quickly and accurately. It hardened in the wink of an eye, thanks to the heat and dryness that prevailed in March in this part of Niger. Erosion from the constant winds threatens the quick ruin of the best footprints. This is why we later came back to this locality, and the two best footprints, already missing a piece when we returned, were placed on a wood frame and brought to the National Museum of Niger in Niamey, where they can still be found.

At that point our team, still led by the geologists from the Atomic Energy Commission, got back on the trail north to reach the well of Azog before nightfall. Under a sky full of constellations of stars, the same ones that the dinosaurs of Irhazer may have contemplated, we set up our camp. Around the campfire, our geological friends told us about what their lives were like, their difficulties, and also their joys of prospecting for uranium in Niger. And our elder statesman, Albert-Félix de Lapparent, told us about his recent discoveries of dinosaur tracks – in France! – just a few months earlier.

It turns out that during the summer of 1963 a chemical engineer, Gilbert Bessonnat, discovered by chance some small tracks that intrigued him, while he was studying the rocks that were visible at low tide on the beach at Veillon in the Vendée, not far from the Sables-d'Olonne. These tracks, impressed into sandy-clayey rock, were presented to de Lapparent, who immediately recognized them as reptilian footprints. They came from a fairly unusual setting, a tidal zone on the strand, and they were visible only at low tide, for several hours at most. During low tide at the spring equinox of March 19, 1965, de Lapparent and his colleagues from the Catholic Institute in Paris carried out a preliminary vis-it and study of the site. They discovered more than a hundred traces of dinosaurs located on two levels in the Early Jurassic strata. The abun-dance, quality of preservation, and the scientific interest of these prints were so impressive that they decided to go back during the next spring equinox in March 1966, and this is why de Lapparent, fresh from the beaches of the Vendée, joined us at Agadès. Still excited by what he had seen, he explained that a single great slab of rock, 20 × 15 meters (65 × 50 ft.), contained dozens of footprints of large tridactyl dinosaurs, and that the total material examined over many hundreds of square meters was climbing toward a thousand prints belonging to at least a dozen dif-ferent kinds of animals. One great slab, six meters long (nearly 20 ft.), had been taken out and reconstructed in the museum of the Sables-d'Olonne, where it can still be admired.

These prints were studied by de Lapparent and his colleagues, who included Christian Montenat, a friend with whom I had studied paleon-tology at the Sorbonne, and Raymond Desparmets, one of their geolo-gy students who came on one of the missions to Niger. They attributed these tracks to the genus *Grallator,* which is thought to be a small, grac-ile theropod, and to the genus *Eubrontes,* which is supposed to be a large carnivorous dinosaur. The Veillon site is particularly interesting, part-ly because it is by far the most important dinosaur footprint locality in France, although there are other great localities, particularly along the edge of the Cévennes. The trackways of the Vendée were particularly difficult to study because of their unusual situation at the edge of the sea. All of the tracks that were susceptible to damage by the action of waves that gradually erode the coast have been removed and are pre-served today in the Museum at the Sables-d'Olonne, in de Lapparent's Geological Institute in the new city of Cergy-Pontoise, and in the Muse-um in Paris. And the work involved in the removal of these tracks by our colleagues was not in vain: The televised announcement of these important discoveries attracted a great number of the curious, as well as inexperienced rockhounds who added to the tidal erosion an even greater human erosion. Today there is not much to see at the beach at Veillon: Between the actions of the ocean and those of people, all trace of these remains has been erased.

Back at the well at Azog, our campfire had gone out. The silence was interrupted from time to time by the bleating of a dromedary or two. I stretched out on the ground in my sleeping bag under a carpet of stars.

The Milky Way looked like a ribbon of torn muslin; shooting stars leapt after one another; and for the first time in my life, I saw the Southern Cross emerge from below the horizon late that night.

Just after dawn, we resumed our work south of Azog on some vast, well-preserved horizontal surfaces, and we were not long in finding about 60 tridactyl prints, followed soon by some new elephantine prints like those at Mount Arli. These were nearly circular and impressed into the ground to a depth of nearly 50 cm (20 in.). The size of the longest of these exceeded 60 cm (24 in.) and had a fore–aft (anteroposterior) diameter of 45 cm (18 in.). At the bottom of these holes the impressions of four digits were clearly visible. On the side opposite to where the digits were impressed, the surface of the ground was pushed up by an enormous lump of mud 10–20 cm high (4–8 in.) that formed a half-circle around the part in front of the impression. This lump was formed by the pressure of the front of the foot as the animal brought it out of the soft mud. The same slab showed tracks with four or five toes, much less deep but as large as the previous ones. Only the front part of the print was well defined. These two types of impressions were made by other large sauropods, who had frequented the same spot after the clay had partly dried out.

At Mount Arli, just as at the site near the well of Azog, the vast slabs littered with dinosaur tracks preserved other traces and prints that allowed us to complete our survey of the milieu and the environment of the area at that time. During the Early Cretaceous, the sediments on the plain of Irhazer were sometimes covered by a thin layer of water. This water, under the action of wind, formed little wavelets similar to those that you can see today at the edge of a lake. Moving rhythmically to die at the edge of the water, these wavelets disturb the sediment at the bottom of the lake. The sediment is then redeposited following the rhythm of the wavelets to assume a shape like a sheet of corrugated metal. These undulations, which we call *ripple marks,* remain and are fossilized after the water evaporates. The presence of ripple marks on these slabs with dinosaur footprints shows us that there could have been a slight temporary current of water at that time. Other portions of these silty tracks bear the impressions of small half-ovals that at first sight look like the shells of tiny clams. Looking at these natural molds under a magnifying glass, we soon discovered that the form of these valves bore the very clear growth marks of conchostracans. These are small, freshwater crustaceans generally measuring less than a centimeter; their bodies disappear completely inside a thin chitinous carapace formed by two connected valves. These comprise a family placed among the branchiopods, freshwater crustaceans with many pairs of limbs and large gill appendages that are known from since Devonian times, 350 million years ago.

This discovery was very interesting because living conchostracans occupy temporary or permanent pools and small stretches of fresh water. They rest in the silt on the bottom, or swim around, feeding on microscopic prey. They live for one or two months on the average. Female conchostracans carry eggs that hatch into segmented larvae if the surrounding conditions are favorable; but if evaporation is too quick, the eggs remain glued inside the walls of the valves, where they can resist long desiccation, to hatch during the next inundation. Desiccation even allows them to be carried by the wind to colonize other more humid and favorable spots. Today, conchostracans are known at all altitudes and in all climates. Their normal habitat is fresh water, and some species can live in brackish water, but none is fully marine.

The study of these conchostracans by my colleagues, Mmes. Defrétin and Guérin-Franiatte, who specialize in these organisms, allowed us to assign the specimens found on the plain of Irhazer to the genus *Euestheria,* which had previously been identified in the Late Jurassic of Bahia and the Early Cretaceous of São Paulo in Brazil, as well as on the African continent, in Cameroon and Zaire. In the Irhazer formation, these conchostracans are found clustered in an irregularly bedded sediment, sometimes made of oblique microstratifications. This seems to indicate a relatively active current, so we concluded that at the beginning of the Cretaceous the plain of Irhazer was a floodplain. This is how paleontologists use paleoecological studies to try to understand the relationships of all the extinct organisms to the surroundings in which they lived.

As I pondered the paleoecology of this area, littered with all sorts of imprints, my imagination began to wander. The tracks of dinosaurs on these great flat surfaces were perfectly preserved; all around the area the countryside was completely flat, with a few meager shrubs and a mat of grass. And even today, the plain of Irhazer is still a floodplain . . . but . . . wait: On the horizon, I was seeing a troop of sauropods, many individuals with their long necks and arched backs; they were grazing, moving along at a quiet pace. The illusion was perfect, but it was only a herd of dromedaries strolling in a *wadi.* For a fraction of a second, I was living in the Cretaceous . . . before realizing that the sun had played a trick on me.

We got back in our vehicles to head north-northeast. The countryside changed progressively, and we left the featureless plain of the clays of Irhazer to cross the sandier underlying beds. We were approaching the western edge of the granitic mountain range of the Aïr. The geologic beds were slightly folded here, following the uplift of this range, and they dipped slightly to the west. As we got closer to this range, the beds that we crossed became geologically older. The countryside grew more chaotic and uneven, and the drivers had a lot of the greatest trouble finding a passageway through the sandy blocks. We had the distinct impression of rolling downstairs in a cardboard box. Our heads knocking the roof and our arms rattling against the doors, but safe in all respects, we reached the edge of a large *wadi* that in the rainy season must carry an impressive amount of sand, clay, and gravel in the rainy season, judging by the innumerable thick, vast sediments that covered its bed. This was the Makarene *anou,* that being the term the Tuaregs use for a *wadi.* The bed of the Makarene *anou* was oriented east–west, and it was a good way to approach the range of the Aïr without crossing the sandstone banks. But the independent axes and the four-wheel drives of our all-terrain vehicles were put to good use on this surface, whose soft ground enveloped the tires like quicksand.

We finally arrived at Aodelbi, a point located 125 km (nearly 80 mi.) north of Agadès. Here the marine limestones of the Carboniferous period are covered by a series of continental deposits whose age is not reliably known. On these limestones rest the sandy clay beds of Moradi, which have been dated to the end of the Permian by some beautiful discoveries of early reptiles. Above these sit the sandstones of Téloua, which begin with a conglomerate of quartz pebbles, many of which bear facets worn down by the wind that give the bed the character of a fossil *reg.* Just 1.2 meters above this conglomerate, Claude Valsardieu, a geological engineer from the Atomic Energy Commission, discovered 12 very nice prints arranged in a most interesting trackway on a beautiful sandstone slab of roughly 1.50 × 1.80 meters (about 5 × 6 ft.). The animal that had left these tracks was quadrupedal, and it had five fingers on its hands and feet. It was *plantigrade,* that is, it walked on its whole foot, as we could see from the impressions of its palms and heels. There was a big difference in size and shape between the hand and foot: The latter was much larger, with the fifth (outermost) toe very well developed and turned outward. This gave the print of the hind foot a very characteristic appearance, because what we would usually call the "lit-

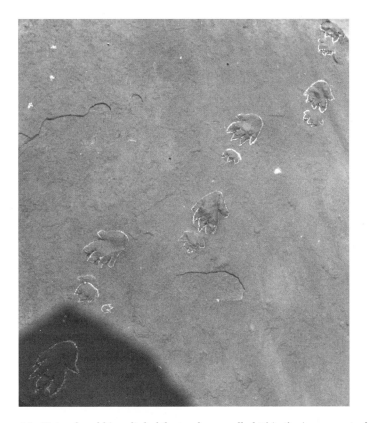

Figure 12. Natural mold in relief of the trackway called *Chirotherium*, near Aodelbi, Niger. It was left by a small quadrupedal Triassic reptile. These footprints are peculiar because the hind foot has a fifth toe shaped just like a thumb. This orientation, which seems counterintuitive, long mystified paleontologists, who wondered how the animal could have walked. (Photo by P. Taquet)

tle toe" was larger than the others in this animal. The size of the digits of the foot increased from the first to the fourth. Similar tracks had been described in other parts of the world and been ascribed to a reptile that was called *Chirotherium*. The predominance of the fourth digit showed that this was a primitive type of *Chirotherium*, and this type of track was found in geological strata that all belong to the Early Triassic of North America and Europe. This was the first time that *Chirotherium* prints had been found in Africa, and the presence of a primitive type of these tracks at the base of the Téloua sandstones, above the beds of the Late Permian, supported a Triassic age for these beds.

Just what kind of animal left these curious tracks on the ground? The footprints of *Chirotherium* were first described in 1833, but it wasn't until 1965 that their trackmaker was identified with reasonable certainty and one of the most troubling puzzles of paleontology was finally resolved. These prints are a bit reminiscent of the shape of a human hand, with their four long digits and a large divergent thumb. They have been observed and collected since the beginning of paleontology in Germany, where they were relatively abundant because of the vast Triassic exposures in that country. In 1835, the German paleontologist J. Kaup proposed the name *Chirotherium* (which means "beast hand" in Greek). The scientists of that time were all very intrigued by the peculiar disposition of the fingers, because the most robust digit, comparable to the thumb, was on the outside of the hand, where the little finger should be, instead of on the inside as it is in all living animals, whether it be a salamander, a lizard, or a bear. And if you get down on all fours, you will easily see that this also applies to Primates – including ourselves.

Was there an animal in the Triassic with its thumbs turned outward? It seemed incredible. Some authors imagined that this wasn't really the impression of a digit, but simply an excrescence of skin on the foot. Others, such as Richard Owen, the man who in 1842 created the concept of the dinosaurs, attributed these mysterious impressions to a labyrinthodont amphibian. (A "labyrinthodont," as they were called, looked like an enormous salamander whose teeth in cross section show a great number of folds in their dentine, the bony substance inside the teeth; hence the name "labyrinth-tooth.") Owen was persuaded that the impression of this large, outwardly turned digit really was that of the thumb, and he concluded that the trackmaker must have walked crossing its hands in front of it! This would have been quite a sight: An animal advancing, putting its left hand to its right side and its right hand to its left side, would probably trip itself and fall down a lot. A reconstruction of a sort of giant frog with its feet crossed was even published in 1855. But most paleoichnologists did not take this seriously.

For more than a century a flurry of hypotheses was advanced; the tracks were variously attributed to a monkey, a bear, a marsupial, an enormous amphibian, a crocodile, and a dinosaur. In 1925 the most complete study of the footprints of *Chirotherium* was published by Wolfgang Sörgel. He showed incontrovertibly that the supposed thumb was really the outermost digit, the fifth, and that feet having this very unusual construction were present in more archaic reptiles considered to include the ancestors of dinosaurs. These reptiles were called "thecodontians" by virtue of their teeth, which were implanted in sockets in the jaws without being cemented in. Sörgel saw that these animals externally resembled crocodiles; but at the time he was writing, no known "thecodontian" could be identified as the *Chirotherium* trackmaker.

It was not until 40 years later that paleontologist Bernard Krebs, then working in Switzerland and now recently retired from a long career in the paleontology department of the Free University of Berlin, described the very well-preserved skeleton of a new kind of "thecodontian" from the Middle Triassic of Switzerland. This archosaur, which Krebs named *Ticinosuchus ferox*, was about 2.5 meters long (8 ft.). It looked a bit like a crocodile, but the construction of its feet was quite different: The organization of its digits and phalanges corresponded perfectly in form and proportions to the trackways of the celebrated *Chirotherium*. Bernard Krebs had finally cracked the mystery of the footprints, with a skill that would have made Hercules Poirot and Sherlock Holmes proud.

Dinosaur footprints had previously fascinated the "father" of Sherlock Holmes, Sir Arthur Conan Doyle, who had the idea of writing what became one of the most celebrated science-fiction novels of all time, *The Lost World* (not to be confused with the Michael Crichton novel from which the sequel to *Jurassic Park* was made). Following the discovery in 1909 of *Iguanodon* footprints in the Early Cretaceous beds of Sussex, south of London – and more precisely, in Crowborough, where Doyle lived – the author of the famous detective novels had the idea of writing a story in which a British paleontological expedition – led by a celebrated personage created for the occasion, Professor George Edward Challenger – set out for an unexplored region of Amazonia and discovered a lost world inhabited by living dinosaurs. Doyle was very much interested in the discovery of dinosaurs right near his home, and he had written to the Museum of Natural History in London to ask an expert to visit. The paleontologist Arthur Smith Woodward came on the scene to authenticate these tridactyl impressions, which he attributed to an *Iguanodon*. Conan Doyle made a mold of the best of these footprints, and used photographs of them to illustrate the cover of the deluxe edition of *The Lost World* in 1912.

I had read *The Lost World* and had been tremendously excited by the story, putting myself in the place of the explorers in Professor Challenger's expedition. For me, these discoveries of dinosaur and reptile footprints on the plains of Niger brought back my childhood dreams:

Suddenly Lord John, who was walking first, halted with uplifted hand.

"Look at this!" said he. "By George, this must be the trail of the father of all birds!"

An enormous three-toed track was imprinted in the soft mud before us. The creature, whatever it was, had crossed the swamp and had passed on into the forest. We all stopped to examine that monstrous spoor. If it were indeed a bird — and what animal could leave such a mark? – its foot was so much larger than an ostrich's that its height upon the same scale must be enormous. Lord John looked eagerly round him and slipped two cartridges into his elephant gun.

"I'll stake my good name as a shikaree," said he, "that the track is a fresh one. The creature has not passed ten minutes. Look how the water is still oozing into that deeper print! By Jove! See, here is the mark of a little one!"

Sure enough, smaller tracks of the same general form were running parallel to the large ones.

"But what do you make of this?" cried Professor Summerlee, triumphantly, pointing to what looked like the huge print of a five-fingered human hand appearing among the three-toed marks.

"Wealden!" cried Challenger, in an ecstasy. "I've seen them in the Wealden clay. It is a creature walking erect upon three-toed feet, and occasionally putting one of its five-fingered fore-paws upon the ground. Not a bird, my dear Roxton – not a bird."

"A beast?"

"No; a reptile – a dinosaur. Nothing else could have left such a track. They puzzled a worthy Sussex doctor some ninety years ago; but who in the world could have hoped – hoped – to have seen a sight like that?"

His words died away into a whisper and we all stood in motionless amazement. Following the tracks, we had left the morass and passed through a screen of brushwood and trees. Beyond was an open glade, and in this were five of the most extraordinary creatures that I have ever seen.

I had just found my *Lost World,* here in the heart of Africa; and over eight years I came to discover, little by little, how to explore it, making its ancient flora and fauna live again and putting my feet into the steps of the dinosaurs. Reality, for me, had largely surpassed fiction.

Eight years, during which the rhythm of one expedition a year pulsed; eight trips to Niger to collect and transport to Paris more than 35 tons of paleontological material, a sediment rich in small vertebrates and dinosaurs, and crocodiles in disarticulated pieces. The help of the CNRS and the National Museum of Natural History was augmented by much other support – first, in 1968, by that of the Fondation de la Vocation, whose president, Marcel Bleustein-Blanchet, and committee members agreed to support me. Three prestigious people warmly supported my application: The first, Francis Perrin, the atomic expert and former administrator of the Atomic Energy Commission, was sensitive to the bonds between basic and applied research, between the research for uranium minerals and the understanding of the conditions under which the sediments that trapped them were deposited. The second, Maître René Floriot, was a talented lawyer, but also, as I learned at that time, a passionate big-game hunter who became very interested in my story of the world's largest crocodile. And finally there was Françoise Giroud, who was then director of the newspaper *L'Express.* I paid her a visit in her director's office where, despite her many obligations, she kindly received me. I was dazzled by the charm and intelligence of this prestigious godmother of public understanding, who swore that she didn't understand anything about dinosaurs but was convinced by the earnestness and the argument of my application.

Then, beginning in 1973 – following the publication in *Le Figaro* of a long article by the great reporter and journalist Pierre de Latil on my research in Niger – a Venetian industrialist, Giancarlo Ligabue, a man passionate about paleontology, and with a degree in the field himself, proposed to finance a part of my expeditions, became my friend, and participated in the 1973 expedition to Niger. This excited him so much that he decided to finance other paleontological expeditions that I proposed – Brazil, Argentina, Saudi Arabia, Bolivia, Ecuador, and Mongolia. All these exceptional journeys permitted us to publish (in Italian, as well as in French, and now in English) a great many works on the natural riches of these countries. These exploits make up much of the rest of this book. Without the support of such farsighted public officials and private philanthropists, we could never pursue the trail of the dinosaurs.

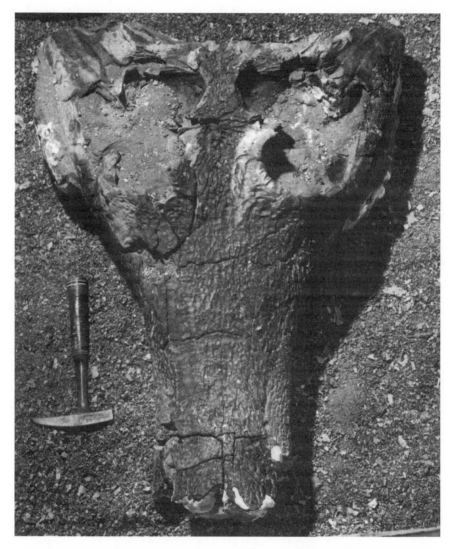

Figure 13. The skull of a giant crocodile called *Sarcosuchus imperator,* from the Gadoufaoua locality, Early Cretaceous of Niger. (Photo by P. Taquet)

CHAPTER FOUR

■

MANY CROCODILES, ONE CONTINENT

THE CROCODILES that live today in the Nile River, and that were considered sacred animals by the Egyptians, have a long paleontological history. Their very distant ancestors roamed the Earth even before the dinosaurs dominated the Mesozoic Era. They diversified into a great number of forms at the end of the Triassic, during the Jurassic, and over the course of the Cretaceous. Moreover, unlike the dinosaurs, they succeeded in surviving the crisis at the end of the Cretaceous and prospered throughout the Tertiary period. Along with the birds, they are the only archosaurs still living today.

During my first expedition to Africa, and through the later expeditions, I encountered some truly unusual forms of fossil crocodiles that helped me to tackle the ancient history of the African continent, and even to follow the phases of its distant origins, in a most unexpected way.

Returning from my first trip to the sands of Gadoufaoua, I stopped off – as requested by the geologists of the Atomic Energy Commission – at the National Museum of Niger, in Niamey, to examine the skull of a fossil crocodile found, a year before my arrival, in the sandstones of a locality I had just visited. This piece, they told me, was exceptional for its great size.

The National Museum of Niger is a unique establishment of its kind. It was created, designed, and realized by a remarkable man, Pablo Toucet, who at the start of his career had been the curator of the museum of Bardo, in Tunis. Pablo Toucet had the idea of creating a new kind of center in the capital of Niger. This center would be a combination of a prehistory museum, a museum of art and popular traditions, a zoolog-

■

ical park, and a display workshop for artisans. Niger is a collection of ethnicities with very different costumes, traditions, ways of life, and languages. Some peoples are nomads, others are sedentary. Hausas, Tuaregs, Peuls, and Kanuris are Nigers who share the same land but mix without really merging. Pablo Toucet had the idea of presenting in a single place the prehistory and history of their country, Niger, to open the doors wide and show everyone the traditional dwellings and the tools of everyday life for each group of native peoples. He wanted to present artisans in the act of creating the traditional objects of their ethnic groups before the very eyes of visitors. On the same site, one group could come to meet and discover others: The Hausa could look into the Tuareg's tent, and the Peul could come into the Hausa's clay house. Under the trees of a great park, traditional music, stories, and legends of each group would be broadcast by a subtle sound system, and small groups of delighted visitors could sit and comment on the customs and costumes of each one. This center, which has no equivalent on the African continent, is naturally graced. It is situated nicely on the edge of the river and it is *the* place in Niamey where visitors converge and a great number of cultural festivals take place. It's also the most restful place in the city, and it is doubtless no accident that it is also the cleanest and best kept in the whole capital. Moreover, in 1965 Pablo Toucet welcomed young orphaned and abandoned children to offer them a trade in craftworking, making carved wooden toys or embroidery, and keeping them away from the risks and consequences of delinquency by giving them some financial opportunity.

In this delightful place, which is held up as an example by all Africans as well as by UNESCO, Pablo Toucet ushered me to the pavilion of prehistory. Here, enthroned on a table, was a gigantic skull of a fossil crocodile with a preserved length of one meter and a posterior breadth of 80 cm. A giant, a monster, whose elongated snout was incomplete, but whose skull must have reached 1.6 meters (about 5 ft.) or longer, this extraordinary animal had lived alongside the dinosaurs of Gadoufaoua; it was certainly new to science, and it seemed to me that it had to be studied.

With Pablo Toucet's agreement, it was decided to ship the crocodile to Paris, to study it and reconstruct the missing end of the snout, before returning it to Niger, where it would be put on permanent display in the National Museum. I got back to Paris, and thanks to the energy of this remarkable cultural dynamo, we set up an operation that brought the cumbersome parcel to the National Museum of Natural History under the best conditions and at the lowest possible cost. Air Afrique kindly agreed to transport it on its Niamey–Paris flight. The skull and its packaging weighed more than 270 kg (600 lb.).

The arrival in Paris on April 30, 1965, of an enormous crocodile did not go unnoticed. The press was there, and the most original account of the event was provided by the famous cartoonist and humorist Alain Saint-Ogan. The front page of the daily newspaper for which he worked showed a man in slippers, very middle-class, sitting in his armchair reading the paper and learning about the discovery of a 100-million-year-old crocodile. The man says to his wife, standing at his side, "And you're trying to tell me that the alligator handbag I bought you last year is already worn out!"

I got to work on the fossil right away. The first question that came to mind was whether there had ever been a crocodile this big on the African plate, or indeed the world. As it turned out, in the course of several missions carried out in the Sahara during the winter of 1946–7, Albert-Félix de Lapparent – yes, the same one – had discovered a number of remains of a large crocodile in the Mesozoic deposits of the northern Sahara: skull fragments, teeth, dermal scutes, and amphicoelous vertebrae (hollowed at both ends, typical of crocodiles). These remains came from the tunnels for the Ben Draou *foggara* near Aoulef in the Tidikelt, that is, in the region surrounding the beautiful oasis of Timimoun. A *foggara* is an artificial subterranean tunnel that conducts water from its source to local towns and farms. Along its length, holes are drilled from the surface to aerate the gallery and evacuate the tunnel. In 1947 de Lapparent announced the discovery of a crocodile, which he called the "Aoulef crocodile," in a note on the stratigraphy and age of the Mesozoic Era deposits in that part of the Northern Sahara. Some additional pieces later came from the same locality, particularly a skull fragment preserved in the personal museum of Commander Augièras at El Golea and some pieces collected by Monsieur Hugot, a schoolteacher at Aoulef. The existence of a giant crocodile in this part of the Saharan Mesozoic was confirmed by the collection of other giant crocodile parts in the extreme south of Tunisia, at the foot of a *gara* – a flat-topped hill or mesa, the Kamboute *gara* – and also during the first explorations of the Gadoufaoua area in 1957 by the geologist Hughes Faure, who collected many large isolated teeth and some dermal scutes not far from the well of El Rhaz.

The study of this collection of material was undertaken by one of my colleagues at the paleontology laboratory of the Museum, France de Broin, who like me was starting her career in paleontology. This unpublished material had been entrusted to her, in the course of her advanced doctoral work, by her uncle! – because France de Broin was the niece of none other than (as you might have guessed) Albert-Félix de Lapparent. The conclusions of her work were very clear: The great Aoulef crocodile, the specimen from southern Tunisia, and the one from Niger all belonged to the same kind of giant, long-snouted crocodile, a new genus. And as luck would have it, France de Broin finished her work just at the moment when the newly discovered, nearly complete skull from the CEA geologists at Niger arrived in Paris. This spectacular and very complete specimen furnished us with all the characters we needed to define the particulars of this new genus, so we decided to prepare a note together that would announce the discovery of this new crocodile to the scientific community. This couldn't have been easier, since we were in the same lab with our desks next to each other.

By comparing the elements of the big Aoulef crocodile to the other known fossil crocs, France de Broin had already shown that this giant form should be placed in what was then called the suborder Mesosuchia. This group of crocodiles was very different than those we know today: Its internal nares (the holes that conduct air from the nostrils to the respiratory system), located on the underside of the skull, opened into the mouth at a position that is much farther forward than in living crocodiles – that is, behind the palatine bones (the back of the hard palate) and in front of the pterygoids – whereas in all recent and living crocodiles, the nares open entirely behind the pterygoids. (In the first crocodiles, the nostrils opened right up in the front of the mouth; thus, during crocodilian evolution, the internal nares have migrated backward in the skull.)

We set to work on our description, and I shall include a paragraph of some particulars of it here – first, because they might be of interest to specialists, and also so that readers who are not specialists can see the kind of detail that's necessary to study fossil anatomy and to establish relationships of organisms. The back end of the skull of this crocodile is large and flattened; its upper temporal openings, that is, the two big oval openings that penetrate the back part of the skull roof and serve as an area of attachment for the jaw muscles, are located very far posteriorly. These openings are broader than they are long, and are very well developed, even larger than the orbits (eye sockets). The frontal bone participates in the border of the temporal opening. The orbits are large, oriented upward, and are bordered in front and on the side by a thick lip of bone. In front of the orbits and on each side of the skull there is a break underscored by a bulge of bone where other crocodiles have a preorbital fossa, or depression. On the snout, the nasal bones contact the premaxillaries in front; the postorbital bones have a prominent bony beak at their external end. The edges of the jaws are straight, not festooned as in other crocodiles. The premaxillary bones are enlarged in front into a spatulate shape and curved ventrally into a prominent beak. There are five premaxillary teeth and 29 or 30 maxillary teeth.

All these characters, added to the position of the internal nares, confirm that this crocodile belongs to the group that was then called Mesosuchians (now, Mesoeusuchians). This animal shared the greatest number of features with another long-snouted mesoeusuchian discovered in the Early Cretaceous of Germany, described in 1841 by Hermann von Meyer under the name *Pholidosaurus schaumburgensis.* But the European species was of very modest size, and in placing our Niger crocodile in the family Pholidosauridae, which was based on the European species, France de Broin and I created a new genus and species whose size made it truly exceptional among crocodiles. We decided to name this giant *Sarcosuchus imperator. Sarcos,* in Greek, means "flesh," and *suchus,* also taken from the Greek, means "crocodile"; *imperator* is meant to emphasize that it was the largest crocodile then known, the emperor of crocodiles.

During the last expeditions to Gadoufaoua, we recovered other skulls and bones of this monster. Another complete skull was 1.6 meters long (5 ft.). Parts of another relatively complete skeleton showed us that the *dermal scutes,* the bony plates that develop under the skin and serve as a sort of armor, could reach sizes of 33 × 15 cm (13 × 6 in.)! These scutes are hollowed out by pits, and this bony ornamentation gives the characteristic design to the skin of crocodiles, which can be seen on the handbags and boots of the well-to-do. The dorsal vertebrae of this crocodile could span 43 cm (17 in.), including the bony transverse processes on the sides, and the diameter of the vertebral spool could reach 12 cm (5 in.). Some teeth, including the roots, reached an incredible length of 14 cm (nearly 6 in.) and a diameter of 4 cm (nearly 2 in.). This was really an impressive animal, and its dimensions would not have shamed any carnivorous dinosaur – even *T. rex.*

I tried to figure out how long this emperor crocodile would have been in life. It's possible to estimate the length of an animal based on the size of its skull, but this measurement can only be approximated, because as the animal grows, its skull and body don't grow at the same rates. We know from having measured living crocodiles that the skull is proportionally longer in juveniles than in adults. Very detailed measurements, taken by Wermuth in 1964, showed that the bodies of long-snouted crocodiles grow at a slower rate than the skulls, compared to the rate in the short-snouted (or *brevirostrine*) crocodiles. Wermuth's observations and tables of measurements allowed me to estimate the length of *Sarcosuchus imperator* adults at around 11 meters (36 ft.).

Eleven meters is a colossal size for a crocodile. The gavial, the long-snouted crocodile that lives today in the Ganges River in India and is considered a sacred crocodile by the inhabitants of that country, can reach respectable lengths. The largest specimen ever killed was measured at 21.5 ft., or about 6.55 meters, and the largest crocodile ever captured was a specimen of the black caiman (*Melanosuchus niger*) whose length reached 23 ft., or 7 meters. The only crocodile that then could be compared to this one is a fossil croc whose remains were found in the beds of the Late Cretaceous of Texas, and was studied by my colleague Edwin H. Colbert. But that crocodile, which was also a giant, is known only from some skull and skeletal fragments, so it is even more difficult to extrapolate its total size. Finally, judging from a portion of the lower jaw of a fossil gavial from the Tertiary beds of the Siwaliks in India, of which a cast is preserved in the Museum in Paris, we can establish the existence of ancestors of our living gavial that must have also attained some impressive dimensions.

The crocodile that was the "Undertaker" of the famous story in Kipling's second *Jungle Book* "was a twenty-four-foot crocodile, cased in what looked like treble-riveted boiler plate, studded and keeled and crested; the yellow points of his upper teeth just overhanging his beautifully fluted lower jaw." *Sarcosuchus imperator* would have surpassed him in length by more than four meters, but their habits were certainly quite different. The anatomical features of *Sarcosuchus*'s snout were, in fact, much like those that we see in the gavial of the Ganges (*Gavialis gangeticus*), which is entirely a fish eater. The long, narrow snout of the gavial, with its uniform teeth running the entire length of the jaws in two long rows, like the teeth of a rake, are perfectly adapted for catching fishes.

But what fishes would have made up the diet for *Sarcosuchus*? In fact, fishes were very abundant at the Gadoufaoua site. There were count-less isolated remains of scales, vertebrae, and skull bones, strewn over the floor of the quarry in such great numbers that we could not avoid walking on a carpet of fossil fish bones. All these fishes were studied by my colleague Sylvie Wenz, whose specialty is Mesozoic fishes. The most abundant of them belonged to the order Selachia (sharks and their rela-tives), which were represented by large, convex dorsal spines ornament-ed in front with grooves and small barbs. It was on these spines that the dorsal fin rested. These calcified pieces were typical of a freshwater shark that bears the name *Hybodus*. Freshwater sharks were very abundant across the entire Saharan platform throughout the Mesozoic Era. Their presence so far from the sea might be explained by these forms swim-ming up rivers, or it might be that they were adapted to a freshwater existence, which is the case today with the freshwater sharks of Lake Nicaragua. But they are also known from many Triassic floodplains in the American Southwest, which were then (as now) very far from the ocean. Along with these shark remains, we also found large, lozenge-shaped scales ornamented on their outer face with a thick coating of enamel. These scales belonged to an actinopterygian (rayfin) fish, more precisely a "holostean," one of an informal group of fishes whose carti-laginous, more or less ossified skeletons were reinforced by these char-acteristic bony plates that articulate with each other. "Holosteans" are represented today by the sturgeon in Europe and the bichir (*Polypterus*) in Africa, as well as by the dogfish of the Mississippi (*Amia*, the bowfin). The Gadoufaoua "holostean" belonged to the genus *Lepidotes*, whose re-mains have been found also in the Early Cretaceous of Europe, in par-ticular in Great Britain and in Belgium at Bernissart, associated with the remains of *Iguanodon. Lepidotes* could reach two meters in length. Among all these remains belonging to fishes were also found some large, flat teeth, striped with a series of crests; these were the teeth of Dipnoans, or lungfishes. These animals can estivate during the dry season, burrow-ing into the sediment and sleeping in a cocoon of clay, slowing down their metabolism and breathing through their lungs. The lungfishes at Gadoufaoua belonged to the genus *Ceratodus*, and they were distant rela-tives of the only Dipnoans who still live today in the rivers and streams of the Southern Hemisphere: *Neoceratodus* in Australia, *Protopterus* in Af-rica, and *Lepidosiren* in South America.

Finally, in 1972 at Gadoufaoua I was lucky enough to discover a site extremely rich in fish remains that represented a lineage of Cretaceous coelacanths. These curious animals, like the lungfishes, had relatives that had played a pivotal role in the conquest of the land by vertebrates some 350 million years ago. The Gadoufaoua coelacanth was represented by two small, complete skulls, and Sylvie Wenz showed that it belonged to a new species of the genus *Mawsonia*. She named it *Mawsonia tegamensis,* alluding to Tegama, the region in which the Gadoufaoua locality is found. The first remains of this large genus were found by Joseph Mawson in the Early Cretaceous beds of the Bahia basin of Brazil; others have since been found in the Early Cretaceous of Morocco. *Mawsonia* is one of the last fossil representatives of the coelacanths. This group was very abundant at the end of the Paleozoic Era, but it was reduced to a single known genus, the British form *Macropoma*, by the Late Cretaceous. Coelacanths seemed to have disappeared since that time because throughout the entire Tertiary, the last 65 million years, there is not a single known fossil coelacanth bone or tooth. This is why the discovery in 1935 of a *living* coelacanth off the Comoro Islands of East Africa caused so much excitement: The lineage was thought to be extinct. The detailed anatomy of this "living fossil," a specimen of which was brought to the laboratory of comparative anatomy at the National Museum of Natural History in Paris, told us a great deal about how the skeleton of these animals changed through time – their fins, their skulls, their vertebral columns. Some of their long-distant relatives, which they share with the lungfishes, had given rise to the first land vertebrates, or tetrapods, probably sometime in the Devonian period.

Freshwater sharks, fishes with big scales, lungfishes, and coelacanths must have made up an abundant and varied diet for *Sarcosuchus.* But in the same place that this giant crocodile was found, another very different form turned up in 1966. And this one was much easier to bring to the lab: Its skull measured only 90 millimeters (mm) in length (4 in.), so you could carry it in your pocket. I had collected this skull and the few bones that were associated with it from a little butte close to our camp, southeast of the well of El Rhaz. We called it "Camp Two-trees" because of the exceptional presence of two acacias close to each other. The pieces that I had collected were encased by a sandstone with calcareous cement; these were not easy to prepare or analyze, but you could see perfectly the anterior part of a very small skull with the lower jaw still connected to it. The pitted ornamentation and the small, conical teeth told us that we were dealing with a crocodile; the rounded shape and the shortness of the muzzle showed that it was a *brevirostrine* or short-faced crocodile. So I transported this precious fossil with all due care to the laboratory. The skull was incomplete at its front end, but it would have been about 12 cm long (5 in.), and the whole animal would have measured about 90 cm (3 ft.).

Starting in 1973, after I finished the description of the new African dinosaur *Ouranosaurus*, I turned over part of the crocodiles that I had discovered in Niger to one of my students, Eric Buffetaut, who wanted to do his advanced (third) thesis and then his doctoral degree on these fossil reptiles. I gave him a part of this material to study, including this little crocodile skull. Plunging into the study of the short-snouted Cretaceous crocodiles of Africa, Eric Buffetaut was led to compare these newly discovered pieces to those that had been discovered previously in Europe and in South America. On a visit to the Natural History Museum in London, he had the opportunity to examine some remains of a crocodile, collected around the turn of the century from the Salvador de Bahia Basin in Brazil, that had been attributed to a short-snouted crocodile. In reality, these Brazilian pieces clearly belonged to an enormous, long-snouted crocodile. They were very similar to the remains of *Sarcosuchus* that I had collected only a few years before in Niger! This unexpected discovery stimulated me to cross the Atlantic Ocean, to visit a continent entirely new to me, and to plunge myself into the enchantment that is Brazil. I had to see how similar the Cretaceous faunas of Brazil and western Africa might have been.

On the Brazilian coast, the most attractive city is without question Bahia, which has kept all the charm of the Brazil of yesterday and was the capital of the empire until 1763. This superb city, with its narrow streets, its many Portuguese-style churches, its flowered houses, its blindingly white walls, and its multicolored population, is unquestionably the most Brazilian part of Brazil. For the geologist – although it may seem less poetic – Bahia is also the place where a vast fault zone stretches out. Following the breakup of the ancient South American shield, a part of the shelf sank between two long fractures, two parallel faults. This geologic accident, which naturally took a respectable number of millions of years, allowed the buildup of impressive amounts of lakebed sediments that were deposited on the continental surface during the course of the Mesozoic Era. The Bahia basin, therefore, is a unique place: It has witnessed not only the Portuguese colonization, but also the colonization of great animals during the Mesozoic, hundreds of millions of years earlier.

Starting in 1860, the English geologist Allport discovered fossils in the sediments that filled in this basin during the Early Cretaceous, 120–140 million years ago. Allport found some crocodile teeth among the vertebrate remains. In 1869, the famous North American dinosaur hunter O. C. Marsh described some of these teeth, creating a new genus that he named *Crocodilus hartii*. Then, at the beginning of the century, Joseph Mawson, "superintendante de l'estrada da ferro da Bahia ao São Francisco" (i.e., railroad supervisor on the Bahia–São Francisco line), collected many vertebrate fossil remains along the roadcut that was being made for a new railroad line. These fossils were described by the English paleontologist Arthur Smith Woodward in the *Bulletin of the Geological Survey of London*. Mawson and Woodward agreed that *Crocodilus hartii* did not in fact belong to the genus *Crocodilus* but to *Goniopholis*, a midsized croc whose remains were relatively well known throughout the Early Cretaceous beds of Europe. In fact, among the pieces that our two authors described was the front end of a mandible, or lower jaw, that was remarkable for its elongation as well as for its great size: It could belong only to a long-snouted crocodile, and so its assignment to *Goniopholis* was incorrect, because the latter is short-snouted.

Thanks to the discovery of many well-preserved bones of *Sarcosuchus imperator*, it became possible to check the fragments from Brazil and the more complete ones from Niger against each other. The comparison was instructive and eloquent. In both cases, the front end of the mandible had the same spatulate shape, with the third and fourth tooth sockets, or alveoli, greatly enlarged. This indicated the presence of powerful teeth on both of the outer front sides of the jaw. Moreover, the dorsal scutes embedded in the skin had an anterolateral spine that was strikingly identical in the two samples. There was no doubt. The African pieces showed that *Crocodilus hartii*, alias *Goniopholis hartii*, was none other than a *Sarcosuchus* which, in the absence of more complete comparisons, given that the Brazilian material is so fragmentary, should be called *Sarcosuchus hartii*. Once it was shown that the two crocodiles belonged to the same genus, the Brazilian croc took the genus name of the African croc, even though the Bahia fossil remains had been discovered first, because they had initially been mistakenly assigned. So the same genus of giant, spatulate-snouted, spiny-scuted crocodile had lived at the same time, the Early Cretaceous, in Niger and Brazil.

But there were further insights into the affinities of these two Mesozoic faunas. Joseph Mawson had also collected in the Bahia basin some

Figure 14. Comparison between the specimens of *Sarcosuchus* found in Brazil (1, 5) and the corresponding parts of *Sarcosuchus imperator* from Gadoufaoua (2, 3, 4) reveals a great resemblance between the jaws (1, 2, 3) and the dorsal scutes, which have a ridge running forward and outward (4 and 5). The mandible shown in 2 belongs to a juvenile, which explains its narrower appearance. (Photo by D. Serrette, MNHN, Paris)

fishes, studied first by Woodward and more recently by our British colleague Colin Patterson. In the continental series of beds from Bahia that comprise the Reconcavo Basin, our railroad superintendent had found the remains of two species of coelacanth and three species of a holostean fish that now carry the names *Mawsonia major, Mawsonia minor, Lepidotes mawsoni, Lepidotes souzai,* and *Lepidotes roxoi. Mawsonia! Lepidotes!* Where have we seen those names before? I had just found them in great numbers at Gadoufaoua, where they had presumably filled the digestive tract of *Sarcosuchus imperator*! Patterson had shown that there were many similarities between the fauna of Reconcavo in Brazil and the fauna of Cocobeach in Gabon, where a coelacanth and a *Lepidotes* had been found. With the big crocodile and the fishes of Gadoufaoua, we could confirm and even bring more evidence of similarities among the faunas of the basins of three very distant areas: Reconcavo, Gabon, and Tegama.

What explains the similarities among these faunas, in areas that are now separated by thousands of kilometers? It was thought for a long time, right up until the 1960s, that the Earth's face had changed little through geologic time. Geologists had concluded that the continental masses were solidly anchored in the depths of the Earth, immobile, and that the continents could experience only risings and sinkings, as during the formation of mountain chains or other large-scale deformations caused by internal movements of the mantle that lay beneath the terrestrial crust. It was thought that these deformations were the main cause of the rise and fall of sea levels: When the crust sank, the seas would advance, or *transgress,* and often flood the continents; and when the crust rose, the seas would roll back, or *regress.* According to the geological community, particularly the geophysicists who studied the physics of our planet, the continents were rigid blocks, and the mountains could never move around and contact each other. However, in 1910 a man of true genius had proposed a theory suggesting that the continents were *not* immobile, and that through geologic time they had drifted apart. This exceptional man was named Alfred Wegener, and he was a talented meteorologist, a geologist, and a peerless explorer, even though he had no degree in geology or geophysics. Born in 1880, he devoted his life to the discovery and the understanding of our planet, right up to his tragic end during an expedition to the Greenland ice sheets, where he died at the age of 50, probably from a heart attack due to overwork.

Looking at the map of the world, Wegener (like many before him) was struck by the complementary shapes of the continents: "Is not the eastern coast of South America precisely adapted to the western coast of Africa, as if they had once been joined? This corresponds even better on the bathymetric map (which shows the curves on the level of the ocean basins) of the Atlantic Ocean, when the contours of the slopes that descend toward the ocean depths, instead of the current coasts, are compared." To these arguments from geography Wegener added others taken from the study of past climates, thanks to a fruitful collaboration with the climatologist Wladimir Köppen, who became his father-in-law. Together they published a book on climates of the past, or *paleoclimates.* In this work they showed very convincingly that the distribution of characteristic evidence of past climates on the continents today could be understood only if the continents had occupied positions relative to the poles and the equator that were much different than they are today. (Consider, for example, the Late Paleozoic moraines – deposits of gravel, rocks, and other sediment left by glaciers during temporary rests as they retreat – that show an ancient glacial episode in India, the southern end of the world having experienced extensive glaciation during the Late Paleozoic; or the great trunks of fossil trees, such as sequoias, that lived long ago in Spitsbergen, Norway, which is now within the Arctic Circle.) What's more, putting outlines of the continents back together, Wegener saw that the distribution of the coal deposits from the Carboniferous period in North America and Europe formed a continuous line, which he thought at the time was at the level of the Equator. The presence of moraines in India was also perfectly explained if it was once joined to South Africa, where Wegener placed the South Pole at that time.

To these arguments, Wegener added many others, taken from geology, paleontology, and the distribution of today's floras and faunas. Wegener could not explain the causes of these movements, and we still don't understand them perfectly today, but his theory was revolutionary. It questioned all the established ideas, and gave the Earth a history that was not static but dynamic. His theory was widely discussed, but frankly did not find broad acceptance at the time. The strongest opponents were the geophysicists, whose equations and models showed that what Wegener proposed was mathematically and physically impossible. There was no mechanism to move the continents through solid rock. The rest of the scientific community – geologists, paleontologists, zoologists, and botanists – added many other criticisms, and after 30 years of discussion his theory was largely dismissed. Wegener died without vindication.

But nevertheless, he was right. It took until the 1950s for physicists and geophysicists who were examining rocks for traces of the history of the Earth's magnetic field to perceive that only continental movement could explain the various orientations of the field through time. When a rock is formed by cooling and crystallization from a magma, or by precipitation from an aqueous solution, the iron oxides in the rock line up parallel to the magnetic field prevalent at the time, like so many tiny compasses. If this orientation is not disturbed later by the heat or pressure of metamorphism, it can be used to tell the direction of the poles at the time the rock was formed. Using this method, researchers soon figured out that all the rocks of the same age on the same continent had the same magnetic orientation, whereas all the rocks of the same age that had been on different continents had orientations that were different from these. However, the tiny magnetic particles in these rocks did not all line up exactly parallel to each other, so lines drawn through their north–south axes did not converge. Assuming that the continents were fixed, one had to admit that the poles had been in several places at the same time, which was impossible. But admitting that continents had moved through time, it became possible to explain the differences in orientation and even to recover the original positions of the continents at the moment that these rocks were formed.

Then, researchers realized that Earth's magnetic field had changed not only position, but direction: The magnetic polarity of the North and South Poles had frequently changed places, so that the poles had flip-flopped, and Earth had lost its compass, so to speak, from time to time. There was no regularity to this phenomenon – it could come at intervals of 500,000 or 600,000 or 800,000 years, for example – so there was much interest in the history of the reversals of Earth's magnetic field, at first on the continents, and later in the oceans. On land, extrusive basalt flows furnished rocks that conveniently registered the succession of magnetic reversals, and these could be dated using radiochronological methods, that is, using the radioactive decay of a potassium isotope (^{40}K) into one of argon (^{40}Ar) in these rocks. This potassium–argon (K–Ar) dating provided the history and succession of all the reversals of the magnetic field back to 3.6 billion years ago.

Finally, using a magnetometer installed on an oceanographic vessel, researchers also began to measure the reversals of the magnetic field in ocean-bottom basalts. Very quickly, American researchers with whom the French scientist Xavier Le Pichon was associated made a discovery that was astonishing and full of implications: The ocean bottoms were organized along great central north–south axes, and on each side of

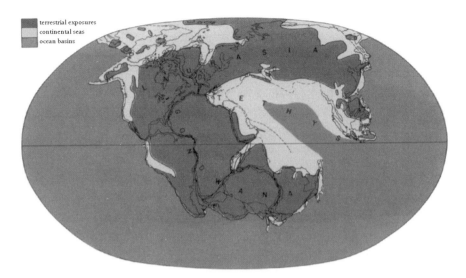

Figure 15. The position of the continents in the Late Triassic. The supercontinent of Pangaea consisted of what eventually became Laurasia (the northern continents) and Gondwana (the southern continents), and this unity explains the relative uniformity of dinosaur faunas around the world at that time. (After Owens 1983, adapted by F. Pilard)

Figure 16. The position of the continents in the Late Cretaceous. The fragmentation of Pangaea into several continents was accompanied by a diversification of dinosaur faunas, which could no longer mingle. Scientists use similarities and differences in these faunas to reconstruct the presence or absence of connections among the various continents at such times. (After Owen 1983, adapted by F. Pilard)

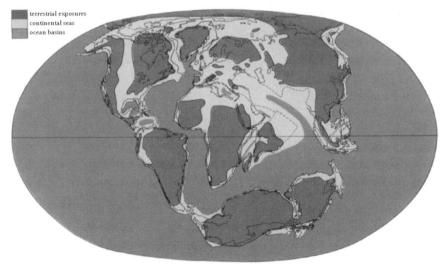

these axes basalts were organized along long parallel bands of unequal breadth that showed alternate patterns of normal (north) and reversed (south) polarity. These important discoveries were correlated with another daring explanation for the formation of the ocean bottoms. Recent detailed mapping of the ocean bottoms, using sonar echolocation, had shown that the oceans were split along their middles by long axial trenches, fractures, valleys out of which, from the Earth's depths, a vast quantity of new basaltic material was continually flowing. On either side of these trenches were long stripes of new basalt, and the farther one moved from the axis, the older was the basalt. Measures of the ages of the different reversals of the magnetic field, and of these basaltic bands, from youngest to oldest, would reveal the age of the formation of these ocean-bottom basalts; we could learn how quickly they were formed, and the direction of their placement. But most important, we could confirm most conclusively that Earth's crust, formed by continental and oceanic plates, was alive with movements, and we could calculate their speeds. The old oceanic material had to give way to the younger material, by being pushed farther from the midocean axes and plunging again into the depths of the Earth. But this didn't happen without complications, including the formation of deep trenches, like off the coast of Japan, and the uplift of immense mountain ranges, like the Rockies of North America and the Andean Cordillera in South America.

Alfred Wegener was right, and the drift of continents and oceans so dear to his heart survives today in the form of *plate tectonics,* as it is now called; his revolution is as central to the earth sciences as the one that Darwin introduced to the life sciences in 1859. Life is not fixed either; it has evolved over millions and millions of years on our planet, which is itself in perpetual evolution.

Today, researchers are engaged in reconstructing the gigantic puzzle of which the present continents are just so many pieces dispersed across Earth's surface. But the contours of these pieces have sometimes become pretty vague, or have changed considerably during their travels as continental plates. This is why paleontology, especially the study of faunas of continental vertebrates and their geographic distribution, helps us to add some important evidence to a good number of pieces in this puzzle, and eventually to recover their ancient associations. With the acceptance of continental drift, the different orientations of magnetites – that is, the iron oxides in basaltic rocks – formed a pattern and found a coherent explanation. With the acceptance of continental drift, the presence

Figure 17. Map showing the three basins of Bahia (Reconcavo), Gabon, and Te-gama (Niger). Clearly the Bahia and Gabon basins were in contact when South America and Africa were joined, as they were here in the Early Cretaceous. (Drawing by F. Pilard, MNHN, Paris)

in these places of very similar extinct faunas and floras – now separated by thousands of kilometers – also formed a pattern and found a coherent explanation. The presence during the Aptian stage of the Early Cretaceous of the same genus of crocodile, *Sarcosuchus,* in Niger and Brazil, associated on each side of the Atlantic Ocean with the same genera of fishes, *Lepidotes* and *Mawsonia,* now has a coherent explanation. If we put South America and Africa back in the positions that they occupied before being separated by drift, we find that the Gabon basin was right across from Bahia; in fact, at that time they formed a single basin. The Tegama basin in Niger, where Gadoufaoua is found, was much closer to Bahia, and some more recent discoveries from Cameroon by a French–American research expedition showed that thick continental sediments of the same age, with comparable faunas, were also laid down in that part of Africa. So there was a faunal unity over a geographic area that formed part of one single supercontinent at the time.

At the end of the Aptian, salt deposits in the Bahia basin attest to the separation that was beginning between South America and Africa. At that time the Gabon–Bahia basin sank and subsided; then the sea transgressed and invaded the continental margins throughout the Albian, 105 million years ago, as shown by the ammonites that have been collected from sediments of that age. Africa and South America separated and left each other for a long period; it wasn't until the eighteenth-century explorations that the native South Americans and West Africans met and formed a new South American–African faunal community!

The faunal community of the Early Cretaceous, revealed by the unpublished discoveries from Niger, would be enriched some years later by the study of the little short-snouted crocodile that I found in 1966 at Gadoufaoua. Preparation of the skull in acid released it from its calcareous matrix, and we could then see that the external nares of this animal were placed very far up along the snout. On each side of the snout, in front of the eyes, there were well-developed depressions. The premaxillary (upper end of the snout) had four teeth and the maxillary behind it had 11, and the third tooth in each bone was enlarged. The teeth in the back of the maxillaries were rounded and must have been powerful crushers. The internal nares, as in *Sarcosuchus,* were placed between the palatines and pterygoids, in a typical mesoeusuchian position. The postcranial material included vertebrae with spoollike bodies, or centra, concave on each end, or *amphicoelous.* These are typical of mesoeusuchian crocodiles, that is, forms like *Sarcosuchus* and living crocs, whereas the first crocodiles had *procoelous* vertebrae, in which the front end of the centrum is concave and the back end convex.

All of the characters of this little crocodile set it among the short-snouted mesoeusuchians, the Notosuchia, which to that point had only been found in South America. The Notosuchia are represented by two families: the Notosuchidae, which lived during the Late Cretaceous, and the Uruguaysuchidae, which lived during the Early Cretaceous. A discovery published in 1959 by my colleague Llewellyn I. Price, an American living in Rio de Janeiro, provided us with some comparisons that had far-reaching implications. Price had described a skull and jaw of a little short-nosed crocodile, which he called *Araripesuchus gomesii,* that came from a locality in the northeast of Brazil in the state of Ceara. I had a vague image of this shrubby, spiny, brush-covered land from the scenery and music of the famous Western-style film by the director Lima Bareito, *O Cangaceiro,* which dealt with the Robin Hood–like bandits of the *sertão,* that impoverished region of Brazil.

The state of Ceara today can pride itself on having one of the richest, if not the richest, paleontological localities of the world. In this region of Brazil, a plateau about 150 km long and 60 km wide, the Chapada do Araripe, is made up in part of the Santana Formation, which is just an immense fossiliferous bed made up of the accumulation of the skeletons of millions of tiny organisms that lived during the Early Cretaceous, at the end of the Aptian, about 110 million years ago. At that time, a very large lake harbored a very diverse flora and fauna. A brutal change in the ecological equilibrium of the habitat destroyed all life, at which point some very unusual conditions of fossilization permitted an exceptional preservation of the fauna and flora that had lived in that environment. The vertebrate fauna includes a multitude of fish species belonging to many different familes, and also of turtles, crocodiles, dinosaurs, and flying reptiles. These are the beds that to this day have yielded the world's greatest diversity of flying reptiles, some of them very strange, with elaborate crests on their skulls. But the insects are equally abundantly represented by magnificent specimens of dragonflies, cicadas, and beetles, and the diversity of the flora is just as interesting.

This immense outcrop has also been farmed for fossils in an unusual way. The fossiliferous layer is found about four or five meters below ground, so the peasants of the Chapada do Araripe dig holes two or three meters across to get down to this level, and then they extract the fossils, which appear in the form of hard nodules in a sandy limestone matrix. These nodules are brought to the surface and immediately split with great skill along their fracture planes. Revealed within the two halves of limestone are two halves of a perfectly preserved fossil. Usually it is a fish, and all the scales, all the skull bones and fins are still in the position that they occupied in the living animal. The fossils of this famous area have been known for a long time, and they are the basis of intense commercial activity. The city of São Paulo is the trading center for these magnificent extinct animals which, apart from their scientific value, are very decorative, and for this reason they are highly prized by collectors of natural curiosities. For dozens of years, the Sunday markets of São Paulo have sold and exported thousands of magnificent fossils, and it must be said that the absence of appropriate legislation in Brazil and the hope of immediate gain have prevented this great country from endowing a museum devoted to the Ceara fossils, a museum that certainly would become in the course of only a few years one of the richest paleontological museums in the world.

Brazil, with its Early Cretaceous vertebrates, its crocodiles from Bahia and Chapada do Araripe, thus became for me a fascinating country, because I could compare its fossil faunas to those that I had discovered in Niger. On January 17, 1977, in a temperature of 30 °C (86 °F) in Rio de Janeiro, I met my Brazilian paleontological colleagues, the fish specialists Monsieur and Madame Silva Santos, the turtle specialist of the Brazilian Mesozoic, Diogenes de Almeida Campos, and Llewellyn I. Price, the author of the little crocodile *Araripesuchus*. The buildings of the Ministry of Mines and Geology of Brazil are found near the base of the famous Sugarloaf, at the summit of which one can admire the superb bay and its magnificent beaches, such as Copacabana. Price was an extremely interesting man, a great connoisseur of the history of paleontology. He had participated in many field expeditions with his North American colleagues, and he had discovered many important fossils, including some early mammal relatives and the first flying reptile from the famous Chapada do Araripe, which he named *Araripesaurus castilhoi* in 1971. When I arrived, Price, like most Brazilians, was on vacation, because it was the austral summer; but Brazilian hospitality is peerless, and he returned to the Institute of Geology just to show me his paleontological treasures.

The contents of some drawers reminded me very much of the fossils from Niger. But these teeth, scales, plates, and bones came from collections made much earlier in Bahia, and since 1940 by Price himself. Some crocodile teeth were exactly like those of *Sarcosuchus imperator,* even though they bore the name *Crocodilus hartii*. The crocodile plates were just as large and had the same ornamentation as those of the giant crocodile from Niger. Other teeth had keels and striations all along their length, and they reminded me very much of the teeth of one of the carnivorous dinosaurs from Gadoufaoua; but here they had been attributed to a long-snouted crocodile named *Thoracosaurus bahiensis* by the American paleontologist O. C. Marsh in 1869. Some pieces of the coelacanth *Mawsonia* were also familiar to me, and some dinosaur vertebrae seemed very similar to those of the sauropod from Gadoufaoua.

I explained to Price that I hoped to go to Salvador de Bahia to look at the geological exposures that were equivalent to the ones in Niger. Price informed me that unfortunately the region is covered by thorny vegetation, and in the absence of any new trails penetrating the brush, it would be very difficult to prospect in the field. But the route from Rio to Bahia was made easy and cheap by the Brazilian airlines, and on January 21 I was in Bahia. I had the fleeting impression of going on a pilgrimage from far-off Africa to find the missing pieces of the puzzle in this city. The unique way of life in Bahia accentuated this feeling. Here, all the churches were full of the faithful who had recovered from illnesses or injuries, and many had come to leave innumerable votive offerings and to hang obsolete crutches and artificial limbs. I thought for a moment what it would seem like to deposit a bony plate or tooth of the giant Niger crocodile on the altar, in the hope of miraculously finding some Brazilian crocodiles.

I finally came to the Portuguese fort of Montserrat, with its ramparts facing Africa and its towers and white walls constructed of beautiful sandstone blocks dating from the Aptian. It was right here that the English geologist Allport collected the first Early Cretaceous fossils from Bahia in 1860. Kneeling on the sands of one of the beaches of Bahia, I was excited to find some pieces of matrix that I thought was very similar to the material I had worked on the other side of the Atlantic. Stretched out on the sand, I could imagine my African and South American crocodiles whirling around in some sort of ungodly romp, a mixture of Brazilian voodoo and Niger possession dance.

Returning to Rio de Janeiro, I found my friend Price and took some photographs of the little skull of his *Araripesuchus*. A visit to their National Museum of Natural History, thanks to the help of another Brazilian colleague, Fausto da Cunha Souza, gave me the chance to examine some of the specimens collected in the previous century by Mawson and some others, particularly the crocodile bones so similar to those of *Sarcosuchus* that are now in the Natural History Museum in London; Mawson's visiting card is still in the drawers. All the pieces, whether of fishes, crocodiles, or turtles, reminded me very much of those of Niger, and this trip to Brazil confirmed for me the idea of a faunal community uniting Africa and South America during the Aptian. Later studies at the laboratory would show that the little short-snouted crocodiles from Brazil and Niger belonged to the same genus, *Araripesuchus*. Together, these animals from the land and fresh water, like *Araripesuchus, Sarco-*

suchus, and fishes such as *Mawsonia,* on both sides of the Atlantic from the same time period, helped to confirm the hypotheses that other pale-ontologists and geologists had proposed: South America and Africa were still united in a single continent until the Aptian, when terrestrial croc-odiles such as the little *Araripesuchus* became unable to cross the strait. Krömmelbein's study of ostracodes, minuscule crustaceans whose tiny shells are found in the basins of Bahia and Gabon, confirmed these ter-restrial connections, because the same species were found on both sides of the Atlantic. Moreover, the study of the marine faunas, especially the ammonites, undertaken by Kennedy and Cooper, showed that the sea did not begin to separate Africa and South America until the Albian, about 105 million years ago. All these facts from paleontology came to-gether to corroborate the facts from the study and dating of the geolog-ic basins by paleomagnetism.

Wegener would probably have been tremendously happy and proud to see his hypothesis of continental drift vindicated by all the earth sci-ences. The waltz of the continents, which sometimes looks like a sam-ba, coherently explains the resemblances and dissimilarities among fau-nas and floras that have succeeded each other through the course of the evolution of our planet. This evolution is not over. The African con-tinent is itself in the process of splitting along its upper reaches. From Djibouti to the great African lakes, Tanganyika and Malawi, a long fis-sure is starting to separate Africa into two subcontinents. Unless there is some change in the program in the history of Earth, the Nile croco-diles, which inhabit all the streams and lakes along this rift, will one day differentiate into populations separated by deeper and deeper, saltier and saltier waters, as oceanic waters invade this nascent arm of the sea from the north. One fine day, some tens of millions of years from now, maybe hundreds of millions of years, an ocean will separate these two fragments of Africa. The Nile crocodiles, or rather their descendants, liv-ing in the west African part will evolve separately from their relatives on the eastern part, and so new species will arise on both sides of this great trench. In some tens of millions of years from now, maybe hun-dreds of millions of years, paleontologists, if there are any left, will look in Quaternary sediments on each side of this new intra-African ocean for the remains of fossil Nile crocodiles that witnessed the ancient uni-ty of this continent.

■

CHAPTER FIVE

IN MOROCCO WITH THE
GIANTS OF THE ATLAS

OROCCO IS a superb land with varied and splendid
countrysides, towns, and villages; its people and their
hospitality seduce visitors from all over the world. Morocco is also a paradise for geologists and paleontologists.
Few countries possess so much testimony to the successive depositions
of Earth's history: Its rock deposits extend from the most ancient Pre-
cambrian times right up to those of the present day.

Absorbed in my research on dinosaurs in Niger, Brazil, Patagonia, and
Madagascar, I'd never had the chance to visit this country, close as it is
to France, even though researchers from our paleontology laboratory at
the Museum had forged strong connections there for years. Along with
the staff of the Service of Mines and Geology of Morocco, my colleague
Jean-Michel Dutuit had carried out some very fruitful fieldwork in the
Triassic beds of southern Morocco, some 215 million years old. They had
brought to light an astonishing array of huge fossil amphibians called
Stegocephalians, which looked a bit like gigantic salamanders.

But one day in 1979 I got a letter from a Swiss geologist, Michel Monbaron, who had been attached to the Service of Mines and Geology for
four years and was in charge of the geologic mapping in the provinces
of Beni Mellal and Azilal, in the heart of the central Moroccan High Atlas. In the spring of that year, Michel Monbaron was starting his fieldwork in the sedimentary basin of Taguelft, east of the city of Beni Mellal. He found a great many fossiliferous outcrops rich in bony splinters
"bleached from gray to white, from three to ten centimeters long, scattered over the surface of the ground; they were clearly fragments of
longer bones that were difficult to identify exactly: ribs, apophyses of

■

vertebrae, or limb bones." From May through July 1978, Michel Monbaron, along with Ahmed Ouazzou and Ahmed Laaroussi, two excellent Moroccan technicians from the Service of Mines and Geology who were trained in paleontological fieldwork by my colleague Dutuit, found and extracted an identifiable bone: a humerus nearly a meter long, evidently belonging to a large quadrupedal dinosaur, a sauropod. Then, working near the Asseksi locality, they discovered a series of dorsal vertebrae whose ribs were still articulated with the backbones. Finally, in August 1978, still in the same region, the little crew found another articulated series of several large vertebrae 30–40 cm (about 12–16 in.) in diameter. All these discoveries, which were unquestionably dinosaurs, stimulated Michel Monbaron to examine the other sedimentary basins in this region. He found a lot of new fossil localities, especially to the south in the Tilougguit basin. Working with the geological authorities of Morocco, and on the advice of Renaud du Dresnay, one of the staff geologists who had lived in Morocco for many years and knew the terrain well, Michel Monbaron invited me to join him and examine his discoveries, to work out their importance, in the hope of getting some more useful information for drawing up his geologic maps.

In October 1979 I went to Rabat and met Michel Monbaron, and we hit it off right away. Originally from the Swiss Jura, he had done his doctoral thesis on the geomorphology of his native land, and had become a specialist on the Mesozoic formations of the Moroccan High Atlas. Three days of preparations at Rabat and our field crew was ready to head southeast. The climb through the foothills of the High Atlas up to the town of Beni Mellal is stunning. Great forests of oak cover large folded mountains of limestone that overhang a plain a thousand meters high. The plain itself was pushed up and carried along on sediments more recent than the uplift of this mountain chain. They say only love can move mountains, but this is really something to see. Today, we know just as surely that tectonic plate movements are the cause not only of the formation and uplift of mountains, but also of their thrusts, faults, folds, gaps, overturns, and lateral displacements that can span many dozens of kilometers. This is all well accepted today, but geologists had some problems swallowing it at first. It was a French geologist, Marcel Bertrand, who showed in 1884, despite the skepticism and incredulity of the geologic community, that the folds in mountain ranges could stretch so far as to become what we call *overthrusts* (reverse faults whose fault planes are more horizontal than vertical), made of ancient materials that

Figure 18. Wawmda, the sauropod dinosaur locality high in the Atlas Mountains, showing the first glimpses of the giant thigh bone as it was being excavated from its Middle Jurassic bed, 160 million years old. (Photo by P. Taquet)

sit atop more recent terrains. And it was in France itself that Bertrand found the proof of his theory. The massive mountain range of Beausset, in the back country of Marseille, is nothing other than a veneer of ancient materials resting on more recent terrains. In Morocco, such folded, uplifted, transported mountains are made of Jurassic and Cretaceous sediments that were deposited at the bottom of an open gulf – to the *east*. That's right: The ocean, 160 million years ago, was not at all where it is today, west of Morocco. The country was still attached to North America. Its shoreline was the east coast, and it bordered a Mediterranean paleoocean called the Tethys, named after the Greek sea goddess of legendary beauty who was the daughter of Uranus, the sky, and Gaia, the Earth.

In the heart of these mountains, the dam at Bin el Ouidane carried a breath of freshness, and the village of Ouaouizaght served as a base camp while we prospected. Michel Monbaron was bent on taking me on foot, on horseback, muleback, and Land Rover to all the fossil sites that he had recorded.

For several days, we surveyed slopes and valleys, rises and hills, ochre limestones and burgundy marls, inspecting all the fossil localities, trying to understand their stratigraphic position and the composition of the fragments of earth we discovered, looking for complete pieces that could help us to figure out where they came from. For a geologist or paleontologist, it is really when you're out rambling and prospecting like this that you feel the earth soaking through all your pores, and you become overwhelmed by the vastness of time and the changes that have followed through the course of millions of years. Curiously, it was an ethnologist who was able to express just what a geologists feel when they're out practicing their trade. But this was no ordinary ethnologist; it was Claude Lévi-Strauss.

Most people don't think of Lévi-Strauss in this context, but he had been fascinated by geology since his youth, and he described in the road notebooks of his famous work *Tristes Tropiques* what it's *really* like to do fieldwork:

[O]n the flank of a limestone terrace in Languedoc, following the line of contact between two geologic beds is something quite different from a hike or a simple survey of the land. This pursuit seems incoherent to the uninitiated observer, but it offers to my eyes the image of knowledge itself, the difficulties that face it, and the joys that can be hoped from it.

Each countryside seems at first to be a huge mass of chaos that leaves you open to choose whatever meaning you may wish to give it. But beyond the vicissitudes of agricultural development, geographic accidents, and the misfortunes of history and prehistory, doesn't the sense of awe inside each of us precede, command, and explain nearly everything else? That pale and blurry line, that imperceptible difference in form and texture of the rocky debris testifies that there, long ago, where today I see dry land, two oceans succeeded one another in time. As I follow the trace of evidence of their eons-long stagnation, broaching all obstacles – sheer walls, landslips, underbrush, gardens – as indifferent to paths as I am to barriers, I seem to be acting irrationally. But the only excuse for this insubordination to the present face of nature is to recover a deeper and more ancient sense of place, certainly obscured today; and compared to it, all other perceptions are only partial or deformed shadows.

Let that epiphany happen, as it sometimes does; sometimes from each side of an inconspicuous crack two green plants of different species surge, each of which has chosen the best ground for itself; and sometimes at the same moment you can see two ammonites in a cliff with differently complicated whorls, attesting in their own way to a gap of tens of thousands of years. Suddenly space and time are confused, even equated. The diversity of life at this moment is juxtaposed, and the past reawakens. Thoughts and feelings attain a new dimension, where every drop of sweat, every movement of a muscle, and every gasp of air become virtual symbols of a vast history. Its movement courses through my body, while my mind grasps its significance. I feel bathed in a deeper understanding, in the depths of which centuries and places respond in terms that finally make sense.

Remembering the truth of this magnificent passage on the flank of these Moroccan mountains, I never suspected that one day I would have the chance to meet Claude Lévi-Strauss personally, and to speak with him – but I did. Not as one geologist to another, but in my capacity as Director of the National Museum of Natural History, to ask him if he would agree to organize an exhibit in the Museum of Man on the objects and the peoples that had symbolized his career as an ethnologist. From this meeting came an exhibit on "The Americas of Claude Lévi-Strauss," which attracted great crowds in 1989.

The first discoveries of Moroccan dinosaurs were made in 1927 by the geologist Henri Termier in the Middle Atlas not far from a locality called El Mers. It was here, in the Bathonian-age beds, about 160 million years ago in the Middle Jurassic, that some years later A.-F. de Lapparent, the French pioneer of dinosaur studies, excavated some bones of a large sauropod dinosaur that he described under the name of *Cetiosaurus mograbiensis*. From his study of these bones, he concluded that this great quadrupedal herbivore was related to a sauropod found in the Jurassic beds of Great Britain. This British herbivorous dinosaur, the first sauropod that was ever described, had been named *Cetiosaurus*, the "whale-lizard," by Richard Owen in 1841. Owen created the concept of the Dinosauria and unveiled it in a lecture on July 30 of that same year to the British Association for the Advancement of Science, in the second part of his work on the fossil reptiles of Great Britain. The genius of Owen was to realize that there had existed in the Mesozoic a group of reptiles that were completely different than those living today. These reptiles had surpassed all known reptiles in size, and had walked with their limbs held under their bodies, like those of mammals. In 1842 Owen baptized them Dinosauria, from the Greek *deinos*, meaning "terrible," and *sauros*, meaning "lizard." Curiously, Owen didn't include *Cetiosaurus* in the dinosaurs at first; he thought it was some kind of giant marine crocodile.

To de Lapparent, the sauropod dinosaur bones discovered at El Mers were quite similar to those of the English cetiosaur, but some differences led him to erect a new species, which he called *mogrebiensis*. Marine beds were intercalated between the beds that held the dinosaur bones, and these beds had ammonite species that indicated a Bathonian (Middle Jurassic) age for the Moroccan cetiosaur – the same age as the English cetiosaur.

In 1942, while doing fieldwork in Morocco, Jacques Bourcart, Albert-Félix de Lapparent, and Henri Termier found the first sparse dinosaur bone fragments near Asseksi in the Atlas of Beni Mellal, where Michel Monbaron was now working. Our three geologists concluded that the Asseksi beds also belonged to the Bathonian stage, and its bones to *Cetiosaurus mogrebiensis*. Later, in the 1950s, other geologists disputed the Middle Jurassic age for these Asseksi beds and assigned them to the Early Cretaceous, despite du Dresnay's reservations. Did Michel Monbaron's new dinosaur discoveries in this area come from the Middle Jurassic or the Early Cretaceous? This was one of the problems that faced my Swiss colleague: Depending on the age of the beds, the geologic history of the region, its deposits, and its tectonic movements wouldn't be the same, and the geologic map that he was to publish wouldn't present the same picture. The problem was far from simple, because a big part of the sedimentary rock deposits of that region consisted of thick series of *red beds* (i.e., beds containing red iron compounds). Were there red beds in the Jurassic, particularly in the Bathonian? Were there also red beds in the Cretaceous? The complexity of the geometry of all these beds in the Middle and the High Atlas made it hard to decide. If we could identify the remains of sauropods that were scattered in the mountains of the High Atlas, we might be able to settle the question. We could at least bring some evidence that could be used in connection with other arguments based on microfossils, tectonics, or absolute dating of the basalt beds that were intercalated throughout the series.

But there was a bigger problem. To a dinosaur hunter like me, cetiosaurs were not about to provide a quick and simple answer, simply because we don't know enough about their anatomy. The British cetiosaur is fragmentary: Only a few pieces of a skeleton were collected, and there was no skull; the proportions between the fore- and hindlimbs are unknown; and the features of the pelvis and feet were not preserved. We know only that the cetiosaurs were among the most primitive sauropods, that their forelimbs seemed to be as long as their hindlimbs, and their vertebrae were massive. We presume that the great Late Jurassic sauropods evolved from these primitive sauropods, branching on one hand into the well-known diplodocids and their relatives (whose forelimbs were shorter than their hindlimbs), and on the other hand into the brachiosaurs and their cousins (whose forelimbs were even longer than their hindlimbs).

In order to be of any help to Michel Monbaron in resolving the problem of the age of the red beds in the Moroccan Atlas, we first had to try to find elements of the skeleton and – well, as long as we're dreaming – find a complete enough skeleton to be able to say something definitive about these great Moroccan sauropods. So, somewhere in these mountains, we needed to find a fossil locality where a number of bones would be coming out on the surface. The bones would have to be well preserved and not too damaged by erosion. This would allow us to excavate the remains with a good chance of finding complete enough elements in articulation – and, if possible, a skeleton. Finally, the deposits covering the fossils shouldn't be too thick, or else we would face a substantial, long, costly, and uncertain excavation. At this point, any experienced paleontologist would simply lean back and say, "Dream on . . ."!

But thanks to his sharp knowledge of the terrain and his talents as a naturalist, in 1978 and the first part of 1979 Michel Monbaron put together a bunch of fossil localities stretching all through the territory that his geologic map encompassed. He saw a clear pattern: Dinosaur bones were not distributed randomly throughout the famous red beds; rather, the richest spots were the "lenses," masses of rock with beveled edges, discolored sediments ranging from beige to greenish-gray, from 10 to 100 meters long (32.8–328 ft.) and about a meter thick, that stood out sharply from the brick red of the rest of the rocks. The sediments in these lenses were clays or silts accompanied by conglomerate layers of gravel. In these lenses, along with dinosaur bones, were often found pieces of fossil wood and little charcoal beds, with fragments of fossil wood colored apple green by their impregnation with malachite or azurite (both copper minerals). It made our prospecting simpler to know that we were more likely to find fossils in places like these.

For two weeks, Michel Monbaron, along with Ahmed Laaroussi and Ahmed Ouazzou, guided me through the North African interior called the *bled*, over great mountains and into beautiful valleys, over abrupt slopes whose folds could be read in all their fullness, to places frequented only by shepherds occupied with their myriads of sheep and goats. We crossed villages perched on the stony ground, where the lumpy cob houses looked like great castles, and where the inhabitants welcomed us kindly with mint tea and took the time to chat. It was harvest time, and on little fields, some seeming no bigger than pocket handkerchiefs, where the ground was covered more by rocks than by wheat stalks, sullen girls were using large sickles to reap a meager harvest. Large threshing areas, well exposed to the wind, were crowded with donkeys and mules, while sacks of grain were carried to small mills powered by the waters of little mountain streams. This pastoral life, though often very harsh – for in the winter snows cover the heights of the Atlas – appears not to have changed for centuries, and time seems to have stopped. For these Berber peasants, we would pass like meteors, being nothing more in their eyes than the prospectors for *madden*, mineral ore.

After examining all the fossil outcrops around Asseksi, Michel Monbaron took me to the area around Tilougguit, which he thought was more promising. We left behind the lake of Bin el Ouidane to cross a steep pass in our all-terrain vehicles. At its summit the cold was sharp, and a brisk wind rushed through the valleys. We had to leave the vehicles near a spring and continue on foot, crossing a vast stretch of eroded limestone outcrops, to reach the locality of Wawmda. In this area my colleague had been able to identify more than a dozen fossil localities, and we spent the day inspecting all of them. But one of them especially caught our eye: Not far from the little stone house of a family of Berbers, on the flank of a small hillock, a great number of dinosaur bones in good condition had been found weathering out of the ground. Spread over a hundred square meters were a dozen vertebrae, including six dorsals in perfect articulation, some bones that looked to me like a sacrum (the vertebrae attached to the hip), and several limb bones including the femur of a large herbivore . . . a sauropod.

A closer look at the layout and the features of these bones indicated that we were dealing with the back end of the animal, and we could make out its pelvis, a hind foot, and the end of the tail. The completeness and the quality of preservation of the material were additional elements in its favor. All the bones appeared to belong to one individual.

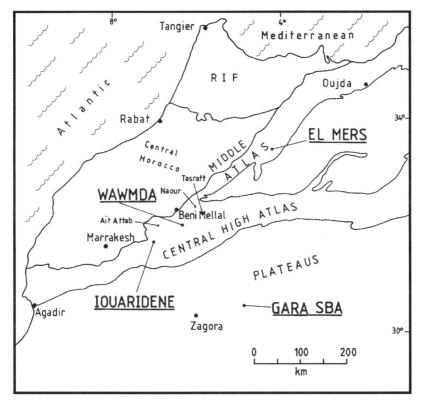

Figure 19. General map of the principal dinosaur localities of Morocco. (After Monbaron 1983)

The disposition of the bones on the ground showed that the front of the animal – its rib cage, neck, and skull – had to be still buried in the little hillside, because the tail vertebrae increased in diameter as they went into the ground; so this showed us the orientation of the animal. Fortunately the hillock itself was not very thick, and we would probably find the rest of the animal inside it without too much difficulty.

I sketched the bones in place, took some photos, and made a rough map of the area. Rain began to fall, and we took shelter in the little house nearby, which was only a few dozen meters away across a small wheatfield. In this modest dwelling, the proprietor, a wrinkled old man, welcomed us warmly, and his steaming mint tea was a delight. A beautiful silk cloth, decorated with motifs, was in progress on a loom. And through the windows we could admire from afar the snow-topped summit of the Rhat, one of the highest mountains of the High Atlas. I was still recovering from the shock that here, in Wawmda, we had encountered one of the giants of the Mesozoic, a sauropod that gave every indication of being in excellent condition. It was just what we'd been looking for. But the question was, would it be enough to help settle our issue about the age of these beds? Moreover, collecting and getting it out would not be a cinch by any means. We had to prepare an excavation plan, collect the supplies, plaster, and tools, find funds to pay the workers, and gauge the amount of time we would need. We made the decision to come back the following year to extract this great dinosaur from its stony bed.

The prospecting tour continued with a visit to the famous dinosaur footprint site of Demnat, worked by the geologists Plateau, Giboulet, and Roch in 1937, where superb tracks of the three-toed feet of theropod dinosaurs were perfectly imprinted on great slabs of red rock. At this place, my colleague Dutuit had found the spectacular trackway of an enormous sauropod in the spring of 1979. Nearly 90 meters long (285 ft.), the trackway showed all the fore- and hind-foot impressions of a sauropod whose stride was nearly 3.3 meters (10 ft.). Its hind-foot tracks were among the largest ever discovered: 115 cm long (45 in.) and 65 cm wide (26 in.). Even if we take into account the fact that the animal was quite probably sliding a bit on soft mud as it moved into its track, which could have lengthened its footprint considerably, we still have to admit that this track is exceptional and that it gives us some precious evidence about the gait of these giants of the Atlas.

Our tour ended when we met up with Michel Monbaron's colleagues Jacques Jenny, Catherine Jenny-Deshusses, and Alain Le Marrec, who were assigned to construct the geologic map of the territory that we had just covered. They had also made some good dinosaur discoveries and wanted to take me out to their field sites. One of these was a superb trackway on a limestone slab, tilted about 45° from the horizontal. The track had been made by a small carnivorous dinosaur, whose fine, slender, three-toed feet were well impressed in the rock surface. The other was even more extraordinary. On the donkey path leading above the houses of Acfarcid, not far from the city of Wazzant, our colleagues had picked up a little vertebra that they had sent me in Paris for identification. This little dorsal vertebra had all the characteristics of belonging to a small carnivorous dinosaur, a theropod, and I was very impatient to get back to the site to see the other pieces that they told me were associated with this vertebra. After an extremely hard climb to the site, I was delighted to find that they really had discovered the remains of a small theropod dinosaur, including the hind foot, part of the vertebral column, and elements of the pelvis! This discovery was even more important because the sediments enclosing the little dinosaur were older than those surrounding the large sauropod bones that we had just examined. These beds were Toarcian, or Early Jurassic, about 185 million years old. Now, we know very little about this time in dinosaurian history, because at that moment the seas were invading the continental surfaces – what geologists call a *transgression* of the oceans. Continental sediments, deposited by rivers, streams, and lakes of that time are very rare, and our understanding of their faunas is correspondingly sketchy. Besides, remains of carnivorous dinosaurs are usually much rarer than those of herbivorous dinosaurs because these Mesozoic predators, much like the lions and cheetahs of the African savanna today, were much less numerous than their prey, as the ecological pyramid would require. My geological colleagues had thus laid their hands on something quite exceptional. Like Michel Monbaron, they had a fine natural instinct, because none of them had ever touched a dinosaur bone before. We collected all the pieces of the little skeleton, even the tiny fragments that had already rolled down among the rock debris on the slope. You can imagine our surprise and delight when we got this material back to the laboratory and found that we had not one but several little dinosaurs at the site! And what's more, the remains of baby dinosaurs were also fossilized among those of others that could be considered adults.

We collected the whole thing, even breaking up the rock of the path itself, to end up with three plaster jackets that we could carry, first by ourselves, and then with the help of a mule, which came in especially handy on the very steep slopes. I'll never forget the surprise on the face of one of the inhabitants of the village, returning home one evening on his mule, who suddenly found a group of people occupied with breaking a big hole in the middle of his daily path, bent on dragging it away to an unknown destination.

This field season was tremendously exciting for me, because I could see how promising it was for the future of dinosaur collecting in the Moroccan High Atlas. So, at the end of the season, we established a scientific research agreement with the Director of Geology for the Ministry of Energy and Mines of the kingdom of Morocco, under the instigation of the authorities Messieurs Ben Saïd and Hilali. Michel Monbaron and

I were pleased and excited, and we began to prepare actively for the next year's fieldwork.

A year later, on September 20, 1980, Michel Monbaron, Ahmed Laaroussi, Ahmed Ouazzou, and I took the path to Wawmda again to relocate our giant of the Atlas, who had been waiting for us on that hillside for 185 million years (or, to be more exact, 185 million and one years). On the evening of September 20, we planted our tents on a pass about a kilometer away from the site, where there was a good spring. My notes of Sunday the 21st read:

Got up 7:30 – Left on foot for the site – ¾ hour walk – The dinosaur is still there – weather overcast – excavation begins – bones all over – Laaroussi took out a tibia – then a scapula and a radius, well preserved – Ouazzou took out a very long rib, then another – it all seems to belong to one animal – Ischium exposed, seems very primitive – a mule brought in our supplies – picnic and tea – freezing weather, with clouds settling on us – Ouazzou built a fire – returned to camp in the evening under a drizzle.

We didn't realize at the time that it would take five months of continuous effort to remove all the bones of this nearly complete, enormous sauropod, one by one. It was five months of unremitting labor by the Moroccan technicians and workers who carried on, in burning sun and freezing cold, in the rain and even snow, extracting the tiny fragile bones and the enormous blocks of this exceptional specimen with chisels, hammers, picks, and awls, brushes, trowels, and needles. It took fewer than ten days in the beginning to take down a meter of the overburden from a surface 80 meters square, to outline the area that was most likely to yield us nearly all the bone. Little by little, a good number of elements of a very massive skeleton were disengaged from their rock coffin. We were soon impressed by the dimensions of these bones. The first one we took out, the right femur, was two meters long. I could lie alongside it and it still surpassed my height of 1.78 meters (5 ft., 10 in.). The tibia was taken out in its turn, and it was 1.10 meters (3 ft., 7 in.) long. Adding in about 40 cm (16 in.) for the length of the metatarsals, which we could see even then, we estimated that the hip socket stood at least 3.50 meters (11 ft., 6 in.) off the ground. What an animal! The series of chest ribs, which were still intact and lying parallel to each other, emerged as the days went on. The largest of these was 2.20 meters long (7 ft., 2 in.). The five vertebrae that were fused to the pelvis, called the sacrum, formed a rigid block 1.5 meters long and 1 meter wide. At the

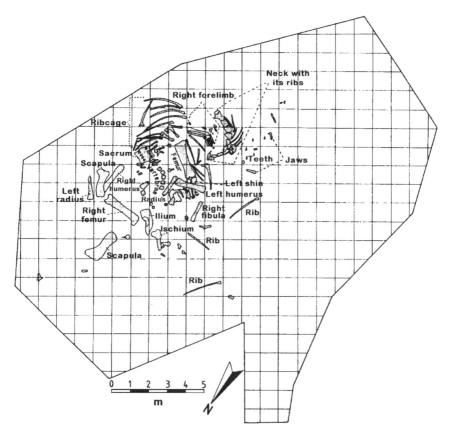

Figure 20. Diagram of the Wawmda sauropod locality in the Moroccan High Atlas. The excavated surface is about 300 square meters. The main bones of the cetiosaur, a large herbivorous sauropod, are shown. (After field sketches by the author, as well as Monbaron 1983)

end of some days of intensive work, a large part of the skeleton was fully visible, and I could begin the initial work of numbering the bones and mapping of the position of each skeletal element.

And through this process, each day, the man who lived nearby and who had offered us his hospitality the year before came to sit at the edge of the excavation, smoking his pipe and knitting wool socks, as Berbers do to prepare for the coming winter. After a while, I began to wonder: What must our charming host have been thinking about us? We had told him the previous year that we would be back and he wasn't surprised to see us again, probably figuring us for ore prospectors. It was unusual enough, to him, that we were putting a huge hole at the end of his wheatfield, but his astonishment grew day by day as these strange objects appeared little by little in the middle of our excavation. To watch the development of totally new questions in the mind of our visitor over those couple of weeks was a unique and fascinating experience. And what a responsibility we were taking on in overturning his routines, his life, and his ways of thinking: We were bringing to light, right before his eyes, a stone's throw from his house . . . bones! Turned into rock! 185 million years old! Belonging to an animal 12 meters long!

Our friend was at first completely incredulous, but after several days he ventured down into the quarry to see for himself the things that we called "bones." The strength of the evidence forced him to admit to himself what he already largely knew from anatomy. Being a shepherd, he knew sheep as well on the inside as on the outside, so he could easily compare the perfectly recognizable ribs of our dinosaur to those of the mammals he knew so well. There was no doubt about it: Even a rib 2.20 meters long (7 ft.) still looks like a rib. A femur is still recognizable, even though it had the respectable length of two meters. But still, how could one accept the ancient age of these objects? Our shepherd took the time to kneel down and smell the bones, to touch them and handle their weight, before forming his opinion. I showed him the plates of illustrations of a work on sauropod dinosaurs, so he could see what we were looking at. But as for explaining to him the evolution of species and the extinction of dinosaurs – well, we all decided to save that for a later date.

The dig continued at a steady pace. Little by little, we prepared the animal's neck, with its long vertebrae and delicate, complicated arches. The neck had been curved backward in an arc by the postmortem contraction of the animal's muscles. By the end of September, the exposed bones offered the impressive spectacle of a dinosaur stretched out in its death pose. At that point we received a visit from the authorities from the Moroccan Service of Mines and Geology, while a Moroccan television crew came to film the unusual excavation. Finally, the governor of the province of Azilal himself, accompanied by the *caïd*s (heads) of the various neighboring villages, appeared in our camp, and in the grand tradition of Moroccan hospitality his visit was marked by a great reception under a superb tent all dressed out for the occasion. The crew members and the visitors shared a sumptuous feast, a *mechoui,* and celebrated this wonderful discovery together.

Several days later I had to leave our Moroccan friends, who continued the dig. Over the next two and a half months all the visible bones were excavated, hardened with solvents, and plastered. I had left the Moroccans with the promise of another *mechoui* if by luck the skull of our cetiosaur was discovered. And sure enough, they found what we had all been hoping for: The back of the skull of the animal was just underneath the first cervical vertebrae, including the two branches of the mandible or lower jaw, a maxilla from the upper jaw, and several very nice spatulate teeth. All the bones that could be plastered in jackets small enough to carry were taken out on muleback as far as the trail, where they could be transferred to vehicles. But the Moroccan mule, the famous *brèle*, could not carry more than 120–150 kg (260–330 lb.), despite its hardiness. Unfortunately, the largest plaster blocks that Ouazzou and Laaroussi had constructed weighed nearly 500 kg (1,100 lb.)! All the largest blocks, including the sacrum and the heavy limb bones, were stacked in place. In December 1980, nearly five tons of fossil material had been extracted from the quarry at Wawmda.

The work was taken up again in May 1981 under the direction of Michel Monbaron and with the assistance of the same group of workers, along with the addition of one of the nearby villagers named Moha, an intelligent and kind man who had quickly learned the techniques of excavation and who had helped excavate the dinosaur at Rabat with Ouazzou and Laaroussi. The surface of the dig now stretched out over three hundred square meters, but some vertebrae and two ribs were still on the docket. On top of this, we had to contend with the extremely important question of how to get the larger pieces to where they could be transported. Thanks to the work of Michel Monbaron and the authorities at the Ministry of Mines and Geology, an elegant, effective, and efficient solution was found to bring the jackets as far as Rabat: The Royal Moroccan Police agreed to put no less than a Super-Puma helicopter at the disposal of our quarriers! Within a few trips, all the heaviest blocks were carried by air as far as the capital. I couldn't help thinking, as I had with our Gadoufaoua material, that once again a representative of the dinosaurs had taken to the air.

In September 1981 I returned to Morocco, and at the Wawmda quarry our crew was digging at the north end, hoping to find the few remaining pieces of our cetiosaur skeleton. This led to the discovery of the second scapula, or shoulder blade. In the end it had taken five months to excavate almost 90 percent of the skeleton of the sauropod. The objective that Michel Monbaron and I had fixed upon had been achieved: This was the first time that a nearly complete dinosaur had been recovered from Morocco; it was the first time that the bones of a nearly complete cetiosaur could be studied; and it was the first time that a skull of one of them had ever been collected.

From the excavation, we could determine a number of details about the circumstances in which this cetiosaur was deposited. The position of the skeleton, the distribution of the vertebral column, limbs and ribs on each side of the skeleton, and its presence in the midst of vast deposits with no other fossils, all indicated that we were dealing with a cadaver that had floated to its eventual resting place at the bottom of a meander in a stream, after a relatively brief transport. It had been partly covered by branches and logs, then buried quickly by silts and fine sands. After being deposited, the bones stayed in place for the most part. Three of the four limbs remained connected to the rest of the skeleton; only the right hindlimb had been dislocated. The skull, far more fragile, had had its bones scattered toward the neck and up to four meters away. The left hindlimb included the femur, the tibia, and the ankle in perfect articulation. The right forelimb, lodged between the neck and the rib cage, although dissociated from its humerus, was extremely interesting to us because the radius, ulna, wrist, and hand bones were all articulated and had not budged since the animal's death. The five parallel metacarpals were associated, and the preservation of these bones was so good that I could tell that our Atlas giant had had a wounded right forefoot: Two of the metacarpals had suffered some shock or illness and had been fused together, gnarled and grizzled at their distal ends. When you realize how difficult it is to find well-preserved skeletons of sauropods, and then how many things we don't know about the structures of the feet of these giant herbivorous dinosaurs, to say nothing of the skulls, you can imagine the interest generated by a skeleton like this. The comparison between the lengths of the forelimbs and hindlimbs showed that the front and hind parts of the animal were carried at about the same height. So we were really dealing with a primitive animal, not with a brachiosaur or diplodocid. This reinforced the idea of the age of these red beds as fairly early in the Jurassic. In fact, the work of Michel Monbaron and other geologists such as André Charrière eventually showed that there were indeed Jurassic red beds, such as where the sauropod was buried, and also other red beds higher in the section than these, which belonged to the Cretaceous. And this explained the past contro-

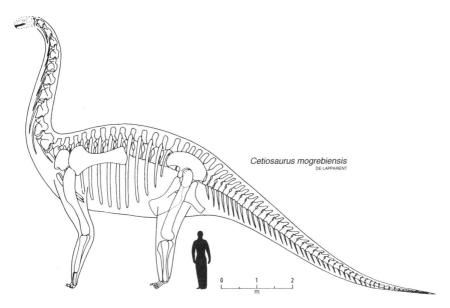

Cetiosaurus mogrebiensis
DE LAPPARENT

Figure 21. Tentative reconstruction of *Cetiosaurus mogrebiensis*. The form of the skull is still hypothetical. (After de Lapparent 1935; Monbaron 1983)

versies about the age of one part of the Mesozoic terrains of the Middle Atlas. The presence of a good portion of the skull pieces – the back of the skull, the quadrate, maxilla, mandibles, jugal – led me to the conclusion that this cetiosaur had a surprisingly long skull, nearly one meter in length – which is quite respectable, and relatively important for a sauropod. In the end our cetiosaur turned out to measure about 15 meters in length (over 50 ft.); the arch of its back reached 4.5 meters off the ground (nearly 15 ft.), and it could raise its head to nearly 10 meters (33 ft.).

A new phase of our work could now begin, with the preparation of the plaster jackets and the separation of all the bones of this enormous skeleton from their matrix. This exacting and time-consuming task would take 24 months at Rabat. It was carried out on the premises of the geological survey by our three comrades Laaroussi, Ouazzou, and Moha. The first two had had long practice in this type of work because they'd already used their talents on the Triassic fossil material that my colleague Dutuit had collected several years before. The results did justice to their superb talents: All the pieces, from the largest to the most delicate, were admirably prepared. Moreover, the large bones such as the limbs and the girdles were relatively fragile, despite their great mass, and each one had to be supported by a cradle of plaster and pig iron to keep it in good shape. Indeed, without these cradles the weight of the bones alone would have caused them to break into thousands of little fragments, and it would have been impossible to reconstruct their original forms and proportions. So each bone was enclosed by two shells of plaster that could faithfully mold each curve of their anatomy, a top shell fitted perfectly to the bottom shell, like a clam or a turtle. These two shells were connected by links that made it safe to transport the great bones; and simply by turning the whole thing over, each side of every enormous bone could be easily manipulated for study.

Meanwhile, as our fieldwork continued, the Minister of Energy and Mines had formed the idea of setting up a museum designed to present the mineralogical, geological, and paleontological riches of their country to the Moroccan and foreign public, and they decided to house this museum in the new quarters that had just been constructed for his Service. The geological and mineralogical part of this task was given to the Germans, and the paleontological part was entrusted to the French, through the scientists at our National Museum of Natural History. The permanent exhibit of this museum was devoted to the evolution of life through geologic time, illustrated by many fossil specimens from the Moroccan collections. This was conceived and carried out beautifully under the direction of Geneviève Meurgues, who was responsible for the museological services of the Museum. The plan was to take advantage of the new and wonderful discovery of a complete Moroccan dinosaur, so they decided to mount a skeleton of our cetiosaur that would greet visitors to the great entry hall of the Museum. This was a new adventure, and to carry it off we needed the help of the French authorities (including the minister of Foreign Affairs), of private companies such as Rhône-Poulenc (who graciously furnished enormous quantities of casting material), of the Moroccan authorities, and of the Swiss Embassy in Morocco. The man in charge of the casting studio in the laboratory of paleontology at the Museum was Jacques Richir, and his son Philippe was also a preparator skilled in casting. Along with the talented Moroccan technicians, they successfully cast the entire skeleton for the hall, following my instructions perfectly. And it indeed became the main attraction of the Museum! The Moroccan post office even based a stamp on it.

During the whole operation I had the great fortune to be able to study the specimen under ideal conditions: I could measure and analyze each piece and prepare a complete description of the animal. All the bones were photographed by Denis Serrette of the Museum's paleontological laboratory, who had come to Rabat especially for the purpose. It was no small task to document this skeleton, because the study of sauropod dinosaurs is a realm in which few paleontologists care to enter. First, there is the problem of handling the bones, which are huge, and you couldn't exactly lay all of them out on a desktop to study them from every angle, even if complete skeletons were available. Then there's the problem that you almost never have a complete skeleton, anyway. And finally, there's the problem that the skeletons are fairly uniform within groups. Certain vertebrae, particularly the anterior dorsals, are useful for distinguishing the genera, and of course it's great if you have skull material. But apart from this, the taxonomy is very difficult.

When I traveled to the United States in 1971, I studied the original specimens of the fabulous sauropod dinosaurs in the greatest museums in America for more than five weeks, in the company of A.-F. de Lapparent himself. I examined the collections of the great dinosaur hunter O. C. Marsh at the Yale Peabody Museum, including the celebrated specimens of *Apatosaurus* (the name now given to the immortal *Brontosaurus*) and *Camarasaurus*. I was hosted by no less a personage than John Ostrom, who was responsible for the discovery that birds had evolved from dinosaurs. Then, at the Carnegie Museum in Pittsburgh, we saw the famous *Diplodocus,* whose collection was financed by the steel magnate Andrew Carnegie. We then got to Utah and its Dinosaur National Monument at Vernal, where a museum provided with a balcony 60 meters long provided a direct view of the excavation of the sauropods *Apatosaurus* and *Camarasaurus* from a rock bed inclined at a 45° angle. Then we crossed the prairies of Wyoming to reach Laramie, where Paul McGrew, a paleontologist and proprietor of a superb mustang ranch, brought us to Como Bluff. This locality was discovered in 1877, not far from the first cross-country track of the Union Pacific Railroad; it delivered the sauropods and other dinosaurs found by Marsh's workers from Como Bluff to the Yale Peabody Museum. Finally, we got the chance to meet up with Jim Jensen, the great dinosaur hunter, in the Colorado desert not far from Grand Junction. At that moment he was excavating a new quarry in which one of the largest known dinosaurs, now called *Supersaurus,* had been discovered; its shoulder girdle was 2.3 meters long (nearly 8 ft.). But the man who incontestably knew these sauropods better than anyone living, who had most completely documented them, and who had devoted his life to them (when he wasn't teaching physics at Wesleyan University, not far from Yale), was John S. McIntosh.

John McIntosh, or "Jack" as he is known to everyone, had the most to teach me about the great herbivores of the Mesozoic. He knew everything about what to study and how to measure it, and about all the mistakes and "short cuts" in the mounts of specimens in American museums (because many exhibits had been assembled from several different skeletons, so one had to use them with caution when making scientific comparisons). Jack had even developed a table of the different criteria and characters to use in identifying skeletal remains of sauropods.

Today, the study of our Moroccan sauropod is on the right track. At some point I still have to go to England to examine the original specimens that Owen used to name the first *Cetiosaurus* in 1841, because even though our Moroccan sauropod is in all probability a cetiosaurid, it's not clear that it belongs to the genus *Cetiosaurus*. The hunt continues, and will be taken up in the land of Sherlock Holmes himself. A complete memoir on the geology and paleontology of the Wawmda area will be published by Michel Monbaron. History will record that our animal, the best known cetiosaurid to date, was discovered by a "Jurassian" – because Michel Monbaron hails from the Swiss Jura, where the Jurassic period got its name! His country also registered one of the rare finds of cetiosaurids outside Great Britain. This was a specimen discovered in the last century by another Swiss geologist, J.-B. Greppin, at Moutiers (again in the Swiss Jura). It consisted of some fragments of long bones, girdles, rib cage, and pelvis, as well as some tail vertebrae. They were described in 1922 by the German paleontologist Friedrich von Huene under the name of *Cetiosauriscus,* and are on exhibit today in the Natural History Museum of Basel.

Since Richard Owen described the first sauropod, these huge and impressive creatures have become among the most popular dinosaurs. Their great size, their long and distended bodies, and their tiny heads perched on hyperelongated necks have fascinated paleontologists as much as they have the public. And surely the *Diplodocus* has contributed as much as any to their continuing popularity. From 1900 to 1902, the American paleontologist Holland collected the bones of a new species of *Diplodocus* from the Late Jurassic beds of Wyoming. The name of this great sauropod alluded to the "double beam" structure of the bony

Figure 22. The mounted skeleton of *Diplodocus carnegiei* in the Museum's pale-ontological gallery in Paris. In front of the Museum workers sit, from left to right, the envoys of the American millionaire Andrew Carnegie, Mr. Coggeshall and Mr. Holland, and Monsieur Marcellin-Boule, who held the chair in paleontology in the Museum.

arches called "chevrons" that are located beneath the vertebral spools about a third of the way back along the tail. The excavation was fund-ed by the millionaire Andrew Carnegie, and the paleontologists had the good sense to name the new species *Diplodocus carnegiei* in his honor. And that seemed to go over well, so they had the courtesy to name the next species that they discovered *Apatosaurus louisae*, after Mrs. Carnegie, whose name was Louise. Andrew Carnegie was so proud of this discovery that he decided to finance the plaster reproduction of several complete casts of the Pittsburgh specimen – which was also good publicity, of course. (But even this specimen was a composite: One of the hands of *Diplodocus* was in fact from *Camarasaurus*, sent by the American Museum in New York!) This was done in short order, and soon the natural history museums in the largest cities in the world were displaying specimens of this dinosaur, 27 meters long (nearly 90 ft.). London, Berlin, Vienna, La Plata, St. Petersburg, Madrid, and Paris received these ponderous gifts and had them mounted, and their inaugurations took place amid all the pomp and circumstance that would testify to the glory of the dinosaurs – and the glory of Carnegie.

The specimen destined for the National Museum of Natural History in Paris was installed in the paleontology gallery, and was officially received on June 15, 1908, at 2:30 in the afternoon, by the president of the Republic himself, Armand Fallières. To be frank, this worthy person was more used to inaugurating chrysanthemum gardens than sauropod skeletons, but all of Paris turned out into the streets around the Jardin des Plantes to see the Important Personages arrive between two rows of National Guardsmen. In the paleontology gallery, Mr. Holland and Mr. Coggeshall, Carnegie's envoys, were flanked by nearly the entire scientific community, waiting for a grand speech. Unfortunately, Fallières was not a grand orator. In fact, he was completely stupefied in the presence of this skeleton 27 meters long, and the only thing he could say was "Quelle queue! Quelle queue!"[1] – much to the great displeasure of the attendant scientists, and much to the joy of the songwriters of the time. The latter lost no time in tearing off a song that was clever enough to survive for some time in the cabaret show at La Lune rousse; some

---

[1] Literally, this expression means "What a tail!" but the word *queue* has several other meanings, including the general region of the derrière and the male reproductive organ. Hence the embarrassment of poor Monsieur Fallières's unintended double-entendre. Another word in French with a somewhat similar sound, *cul*, has similar anatomical connotations but is considered more vulgar. So, facing a long wait in a line (another meaning of the word) in Paris, one might be careful in pronouncing to one's friends, "Quelle queue!" The French expression "Oh, quelle cul t'as!" means, loosely translated, "Wow, what a nice ass you have!" (A homophonic approximation of that expression, *Oh, Calcutta!*, was the title of Kenneth Tynan's 1969 erotic musical revue, in accordance with its degree of nudity. The 1946 painting *Oh, Calcutta! Calcutta!*, a revealingly draped female nude by the French surrealist Clovis Trouille, was referenced by that production.)

of the verses described the president of the Republic's inability to pronounce the word *Diplodocus*.[2] The event was a great to-do, and a banquet was given at the Museum. Its menu, specially printed for the occasion, offered a series of courses alluding to paleontology: The entrée was served with little peas "à la Holland." Mr. Holland received the cross of an officer of the Legion of Honor, and Mr. Coggeshall received the decoration of an Officer of Public Instruction.

Henry Bidou, a reporter from the *Journal des debats politiques et littéraires,* wrote to commemorate the event on June 16, 1908, in a very evolutionistic article entitled *"Diplodocus* tells his side of the story," in which our celebrated reptile, following Holland and Fallières, took the stage and made his own speech, in a strong neo-Jurassic accent, concluding as follows:

"I am happy to see the light of day again, blinding though it is after so many years. How odd you are, so small and yet so far along the evolutionary scale! I can see that the hair of *Homo sapiens* is regressing; your teeth aren't good for much; your claws are matching sets, but variously colored; and your muscles are atrophying. Soon your successor in the chain of life will appear: a hideous creature with an all-powerful brain, a soft and shiny hydrocephalic who will understand the laws of the universe well enough to supplant you in it. So, ladies and gentlemen, when you are reduced to this sad state of an extinct species, maybe some of you will in turn be preserved in this hall, the pantheon of nature. And I will dearly look forward to recalling sometimes the unforgettable memory of this beautiful and patriotic ceremony, when we renew our conversation here on lonely winter evenings." The monster closed its hideous mouth; the National Guard began to play the "Battle Hymn of the Early Cretaceous"; and its dissonances awakened the President of the Republic, who had been snoring softly.

The architecture of a sauropod skeleton is obviously a sobering subject, and we can only wonder about many questions concerning the physiology and mode of life of an animal with that kind of construction. The massiveness of its limb bones led the paleontologists who first worked on them to conclude that they were amphibious animals; their

---

[2] Especially since the typical Frenchman would pronounce *Diplodocus* (in English phonetics) as something like *Dee-ploh-doh-QUE* – so the verses of the song emphasize that the president can only remember the first three syllables, and forgets the last one, which sounds like *queue* (the only thing he could think of to say when he actually saw the animal).

limbs and muscles, it was thought, would be incapable of sustaining the animal's huge bulk on land. On the other hand, given Archimedes' law, there was no reason why very large vertebrates couldn't live in the water, and the researchers of the nineteenth century who studied the first known remains of these enormous extinct reptiles could compare them only to whales, whose immense masses didn't rule out swimming or diving. This is why the first sauropod was named "cetiosaur," or "whale-lizard." This idea was reinforced by the discoveries of some rare skull material of sauropods that had their external nares, or nostrils, located in an elevated position on the top of the skull. It was thought that this allowed them to continue to breathe while the rest of them was submerged. A German researcher, Wilfarth, even thought that sauropods lived completely immersed, floating at the level of their natural buoyancy, using their tails as anchors to resist currents.

However, as more and different sauropods were discovered over the years, the sedimentary contexts of their remains, the floras and faunas associated with them, and the growing understanding of their footprints and trackways as well as their anatomy, all led researchers to bring the sauropods out of their supposedly aquatic habitats.

In 1971, a young and brilliant American paleontologist, Robert Bakker, a student of John Ostrom, began to play a decisive role in giving a new image to dinosaurs and particularly to sauropods. I had met Bob Bakker at Yale when I visited John Ostrom in the middle of my work on the *Ouranosaurus* from Niger; I was comparing my animal with American ornithopods such as *Tenontosaurus*, which John was studying at the time. Bob was working in a corner of the laboratory, bent over a binocular microscope, and I had immediately taken him for a student because he wore his hair very long, like a superb horse's tail, as the fashion was at the time (and he still does, of course). In an explosive article published that same year, Bakker made the hairs curl on many a paleontologist's neck by questioning the aquatic mode of life of sauropods.

An excellent artist, he had previously worked up a completely new drawing of the small carnivorous dinosaur *Deinonychus*, which John Ostrom had studied. Following Ostrom's insights, Bakker had portrayed this predator not as a clumsy, apathetic oaf but, in conformity with its anatomical features, as an alert, graceful animal that evoked a cheetah more than a sluggish lizard. When he began to become interested in sauropods, Bakker showed convincingly that the solid, columnar limbs, the broad, flat feet, the heart-shaped ribcage, the flattened form of the

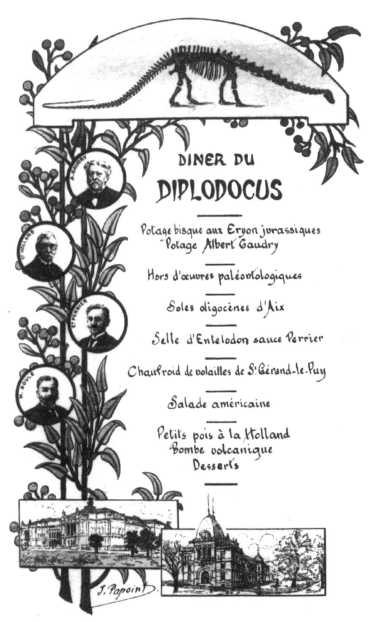

DINER DU

DIPLODOCUS

Potage bisque aux Eryon jurassiques
Potage Albert Gaudry

Hors d'œuvres paléontologiques

Soles oligocènes d'Aix

Selle d'Entelodon sauce Perrier

Chaufroid de volailles de St Gérand-le-Puy

Salade américaine

Petits pois à la Holland
Bombe volcanique
Desserts

J. Papoin

Figure 23. Menu for the dinner in honor of the *Diplodocus* skeleton, which was inaugurated by French President Armand Fallières in the paleontological gallery of the National Museum of Natural History, Paris, in June 1908. Mr. Holland's picture is arranged along with those of the luminaries of the Museum. The inventive names of the dishes on the menu don't give us much sense of their exact contents, but they are full of puns and allusions to fossil animals and paleontologists of the time.

ribs, and the reinforced structures of the backbone, with strong apophyses and powerful spines, were good indicators of animals that were perfectly adapted to terrestrial life. Moreover, the raised position of the nostrils was also found in other terrestrial animals – for instance, elephants, to take an example from living creatures. On balance, the evidence indicated that the great sauropods were terrestrial animals, consummately adapted to a terrestrial existence, and their skeletal architecture was highly functional. Four very solid and vertical columns held up superstructures that could be compared to the struts of a metal bridge like those that Eiffel constructed – or, of course, his famous tower, which also has four legs. These provided the maximum support with the minimum weight. Moreover, in sauropods the vertebrae were lightened as much as possible: Their spools were hollowed out and excavated with side cavities called *pleurocoels,* whereas the struts designed to absorb tensions and pressures were reinforced. The great length of the neck supported a tiny head, but it was not designed to be a snorkel. Instead, it allowed the sauropods to browse in high foliage, like giraffes do today on the savanna, perhaps separating their breathing from their eating. This would have been practical for animals that had to fuel such large bodies; and it gave them a high observation post from which to detect their predators, the theropod dinosaurs.

Bakker then developed his ideas on sauropod ecology. He used a series of arguments taken from the structure of the bony tissues (which, as our colleagues Armand de Ricqlès and Robin Reid have independently shown, look much more like those of mammals than those of typical reptiles); the skull anatomy (including some of the particulars discussed above); and his estimates of prey–predator ratios compared to those of reptile and mammal communities of today. (A lion eats ten times as much as a crocodile because of its physiology, so a community of warm-blooded animals supports far fewer predators.) Bakker concluded that dinosaurs, including sauropods, were warm-blooded: They had a thermoregulatory regime comparable to those of mammals, whereas classic reptiles such as crocodiles, turtles, and lizards had cold blood, a lower and more variable internal temperature that depends on the ambient temperature.

A somewhat euphoric period followed during which "hot-blooded" sauropods were often represented gamboling like goats, their necks held vertically, balanced on their back legs to reach the highest branches of trees. This fashion also faded in its turn, and new work has tempered how we view these dinosaurs. But Bakker's views were instrumental to the "Dinosaur Renaissance," to use the title of his well-known article of 1975 in *Scientific American*. They stimulated new work by paleontologists and contributed to the further popularization of dinosaurs. They had some lasting effects: After our having first represented dinosaurs, particularly sauropods, as large, stupid, and ugly, and then imagining them like mammals living today on the African savanna, we now have ideas that are probably a little better tested and informed. Sauropods were terrestrial animals, well adapted to their environment, living in large herds, capable of making a living on the vast swampy plains of the Mesozoic, and occupying the ecological adaptive zones occupied today by elephants, though they were much larger. Sauropods may or may not have been warm-blooded, but their great mass undoubtedly protected them from rapid fluctuations in body temperature by assuring them of a relatively steady core temperature.

Sauropods, like other dinosaurs, can't be equated with mammals; but they also can't be confused with other reptiles, from whom they differ not only in size but in the vertical posture of their limbs. Clearly, dinosaurs were . . . well, dinosaurs. And this is what makes them unique. Their skeletons, their muscles, their physiologies, and their modes of life allowed them to survive the entire Mesozoic Era, which isn't bad at all considering the longevities of other groups of animals over the course of geologic time.

Some years ago, a man came to see me at the lab in Paris; he was in charge of public relations for the French National Management Center (CNPF), and he wanted to take a picture of our *Diplodocus*. The organization was getting ready to hold its annual meeting in Marseilles and wanted to start it off with a series of talks that would motivate and energize its executives. The CNPF had decided to use dinosaurs a as symbolic counterexample: big stupid dinosaurs, who had disappeared from the planet because they could no longer adapt. I gave him a nice slide of a sauropod, asking if he would kindly tell the CNPF members that dinosaurs populated Earth for 155 million years before disappearing, and that I wished the French executives a similarly long reign.

Figure 24. Skeleton of *Tarbosaurus*, a carnivorous dinosaur very closely related to *Tyrannosaurus*, in the Nemegt Valley, Late Cretaceous, Gobi Desert, Outer Mongolia. (Photo by P. Taquet)

CHAPTER SIX

■

IN THE STEPPES OF
CENTRAL ASIA

L IKE EVERYONE ELSE, Europeans have always been fascinated by Mongolia, a far-off country of steppes, deserts, and men
who spend their lives on horseback. Paleontologists share the
same fascination. But there's something else: We're attracted by
the famous fossil sites of the Gobi Desert, which have been known
since 1922.

The recent political revolutions in the former Soviet Union also
brought corresponding revolutionary changes in Western relations with
Mongolia. Our Mongolian colleagues now have the opportunity to share
their paleontological heritage with the rest of the world, and Western
researchers now have the opportunity to visit this country of which they
have dreamed for so long. So, in 1990, when the Mongolian Embassy
in Paris began discussions with our Museum, whose directorship I had
just assumed, I didn't want to waste any time. We quickly drew up a
plan to exhibit the paleontological treasures of the National Museum
of Mongolia, not just to the French public but to all of Europe. I brought
the plan to my friend Giancarlo Ligabue, president of the foundation
that bears his name, with the idea of bringing this great exhibit of Mongolian dinosaurs and mammals to his home city of Venice as well. But
to make the exhibit as interesting as possible, and to ensure that the
skeletons were reconstructed accurately, we'd want to know as much
as possible about the sites themselves, of course, and we'd want to capture them on film. In presenting the dinosaurs and mammals of the
Gobi Desert to western Europe for the first time, we wanted to exhibit
real, exceptional specimens, complete and magnificently fossilized, to
show their geological context and the grandeur of their surroundings.

■

123

Plus, we could benefit from the experiences of two of our own paleontologists, Denise Sigogneau-Russell and Don Russell, renowned specialists in fossil mammals, to present these small, intelligent creatures who lived at the same time as the dinosaurs, but are usually overlooked by the public.

Fortunately, I knew the man responsible for the geology and paleontology of Mongolia, because I had met him at a conference in Warsaw. This was Rinchen Barsbold, a specialist on carnivorous dinosaurs who had worked extensively on the Mongolian material. Having made our decision, we first took a brief trip to Ulan Bator, the Mongolian capital, in October 1990, to meet with our colleagues and to plan our fieldwork in the Gobi Desert. I took along Bernard Battail, one of our paleontologists who specializes in reptiles and is also a great student of Russian and Mongolian languages. Our contacts with the Mongolian authorities were excellent, and so we planned our fieldwork for July 1991, the best time for going out onto the Asian steppes and the Gobi Desert, where the temperature can plummet to $-30\ °C$ ($-22\ °F$) in the winter and soar to over $40\ °C$ ($104\ °F$) in the summer.

The first contact between the West and Central Asia, particularly Mongolia, took place in the thirteenth century. Genghis Khan died in 1227, and his son Ogoday commanded the troops that overran Europe as far as the Dalmatian coast, on the Adriatic near Venice. To establish ties with the Mongols – and to try to find the fabled kingdom of legendary twelfth-century Christian king "Prester John," a mythical land thought to be located in the heart of Asia – Pope Innocent IV in 1245 dispatched to meet the Mongolian princes the first Christian ambassadors ever to set foot on the plains of Central Asia. Brother John of the Carpini Plain of Italy, originally from Umbria, left Lyon to disappear into the vast, almost unknown wilderness of the Mongolian steppes. He was followed in 1252, at the behest of Louis IX, by the Flemish Franciscan priest Willem van Ruysbroeck.

As a result of these travels, the West was able to establish traditional trading routes with Asia, including the Silk Route. Passing through the great oases of Turkestan, they eventually reached Karakorum. Their accounts convey the extraordinary impression that these immense landscapes made on them. What was the steppe? For them, "A vast desert of thirty days' travel stretching northward where there are neither mountains nor rocks, but grass in abundance." "The desert stretches out like an ocean," Ruysbroeck wrote – but he appreciated the fermented mare's milk, which he compared to the *râpé* wine of Champagne.[1] We can imagine Rubrouck's surprise when he reached Karakorum in 1254 and found a horde of Mongols, Russians, Georgians, and Armenians, as well as a certain Guillaume Buchier, formerly a native of Paris and now the official silversmith of the Khan. All these men had been captured or hijacked by the Mongols and brought to the heart of Central Asia. But Europeans would get no real firsthand knowledge of Asia until 1271, when a man dressed in the traditional clothes of a Tartar returned to his native Venice, which he had left at the age of 17. Marco Polo enchanted, surprised, and fascinated the Western world with the story of his travels, *The Book of Marvels*. His voyage to Peking via the overland route through northern Tibet, and his return by sea in the company of a Mongolian princess, brought to light a great many new and strange facts that made a very powerful impression in the West.

The other route to Mongolia was the one used by the Cossacks to cross the basins of the great Siberian rivers; it was marked by outposts of the Russian mail as far as Lake Baikal. The Russian Nikolas Mikhailovitch Prjewalski was one of the great latter-day Asian explorers. He took his first trip in 1871, crossing the Gobi Desert, turning northeast through Mongolia, and winding up in Peking. Over four voyages he explored many mountain ranges of Central Asia, accompanied by a Mongolian guide, Chuton Zamba. He drew up a great number of maps and brought back to St. Petersburg numerous samples of the floras and faunas that he found on his travels. He described the wild Mongolian horse that now bears his name (Prjewalski's horse), and from which our modern horse may be partly derived.

At the beginning of the twentieth century, the famous American paleontologist Henry Fairfield Osborn, who was in charge of the fossil mammal collections at the American Museum of Natural History in New York, put forth the idea that Central Asia was probably the center of origin of many groups of mammals, including humans. This idea was taken up by William Diller Matthew, who developed the hypothesis in his famous work *Climate and Evolution*, and by Roy Chapman Andrews, both of whom worked at the American Museum.

---

[1] This is a wine made by adding water to what we call in English "grape pomace," the pressings of grapes and stalks used to make other wine, as well as vinegar. Thus it does not connote a wine of the highest quality.

Their first expedition was in the spring of 1922. Twenty-six people participated, including Walter Granger, who was something of an experimental paleontologist and had already been fortunate enough to carry out six seasons of fieldwork in the famous dinosaur beds of Wyoming. For the American researchers, the main problem was to understand the geological and paleontological history of Central Asia, and to find out whether it could have been the center of origin for many mammalian groups, including humans. So they had to bring along some highly qualified specialists on Central Asia. In fact, this was the first expedition of such scope ever to use these methods. The fossil history of Central Asia was completely unknown at the time. In China, fossil bones and teeth had long been sold as medicines, and few of these specimens had been studied by British and German paleontologists. The only known vertebrate fossil from the Central Asian plateau was a sole rhinoceros tooth, discovered by the Russian explorer Obruchev between 1894 and 1896.

Andrews listed four main reasons why Mongolia had not been more intensively explored by Western societies in earlier years:

First, Mongolia is isolated in the heart of a continent, and, until recently, a considerable journey was required even to reach its borders. . . . Second, the distances are great and transportation slow. Mongolia is roughly two thousand miles in length from east to west and twelve hundred miles in breadth from north to south. In all this vast area there is not [in 1922] a single mile of railroad. . . . Third, the climate is very severe . . . the plateau is swept by bitter winds from the Arctic Ocean . . . and bare existence demands the strongest constitution. Fourth, in the Gobi, which occupies a large part of Mongolia, food and water are scarce, and the region is so inhospitable that there are very few inhabitants. . . . I felt certain that all the physical difficulties could be solved by some means of rapid transportation and that without it an expedition such as we had in mind could not be carried on successfully. I believed that the automobile was the answer to the problem. With motors we could go into the desert as soon as the heavy snows had disappeared, penetrate to the farthest reaches of Mongolia, and return before continued cold and snow set in. . . . From previous experience, I believed that in cars we could travel about one hundred miles in a day; that is, ten times as far as a camel caravan. Thus we should be able to do approximately ten years of work in one season. As results proved, we did maintain just about that ratio. Moreover, our investigations were made much more effective than they could have been otherwise, for the geologists and paleontologists were able to leave the main fleet at any time to examine outcrops or exposures which were fifteen or twenty miles away. With camels this would have been slow and laborious.

The expedition of 1925 included 125 camels carrying four thousand gallons of gasoline, a hundred gallons of oil, three tons of wheat, a ton and a half of rice, and other goods in proportion to these. The caravan of camels left gasoline and food at two depots and waited for the explorers near a well eight hundred miles into the the the heart of the desert.

The American expedition that left China in 1922, heading toward Ulan Bator, indeed discovered fossil bones – not of humans, but of dinosaurs. They made a few quick collections, then reached the capital of Mongolia, only to head west toward China, across the Gobi. On the way back, the expedition stopped at the foot of the magnificent red sandstone cliffs called Bayn Dzak. These cliffs were lit up so beautifully at each sunset that the Americans called them the Flaming Cliffs. This is where the crew members found the bones and eggs that they at first thought belonged to birds.

A second expedition was launched in 1923, and excavations were set up at Iren Dabasu and Bayn Dzak. They found some superb skeletons of a dinosaur that was completely new to science, a primitive horned dinosaur that was called *Protoceratops andrewsi* in honor of Andrews, the leader of the expedition. And for the first time, the specimens represented many different stages in the growth series of the species, as well as eggs clustered in nests. Granger and his colleagues concluded that they belonged to the little *Protoceratops.* A third expedition brought back even more fossil riches, and today they can still be seen in the exhibit halls of the American Museum in New York. The results of the expedition were published in 1932, and from that time onward, the Gobi Desert became a legendary place in the annals of dinosaurian paleontology.

Russian paleontologists took over the site after 1945. From 1946 to 1949, a series of very important expeditions was organized by the Academy of Sciences of the Soviet Union, using heavy supply trucks and light all-terrain vehicles. The expedition leader was the Soviet academician Yuri Orlov, and the more famous participants included I. A. Efremov (who wrote excellent science fiction as a sideline) and A. Rozhdestvensky, who related his experiences hunting dinosaurs in the Gobi in a very amusing book. These expeditions were crowned with success, and extraordinary discoveries were made in the western part of the Gobi Desert. The Russian paleontologists collected eggs and skeletons of *Protoceratops* at Bayn Dzak, as well as material from a new kind of armored dinosaur, *Syrmosaurus.* But their most important finds came from the Nemegt Valley, farther west, where complete skeletons of a large carnivorous dinosaur, very similar to the famous *Tyrannosaurus* of North America, were discovered and given the name *Tarbosaurus*. At the same time they were pulling giant duckbilled dinosaurs, or hadrosaurs, out of their rock matrix right nearby. These turned out to belong to a species of the genus *Saurolophus*, which was already known from North America. Other discoveries added to these, notably of ostrich dinosaurs, or ornithomimids. In all, 120 tons of new material, including 10 complete skeletons of carnivorous and duckbilled dinosaurs packed in 460 boxes, were shipped on the trans-Siberian railroad to Ulan Bator. This material is on display today in the magnificent paleontological museum in Moscow, with some on exhibit at the museum in Ulan Bator.

From 1952 to the present day, our Asian colleagues have published a great many articles and monographs describing the extraordinary fauna of the Gobi. Rozhdestvensky, Chuvalov, Tumanova, Maleev, and Kurzanov have shown us a whole flock of new dinosaurs: the great *Tarbosaurus* (now thought to be so close to *Tyrannosaurus* that it has been put in the same genus), the enigmatic *Avimimus*, the armored ankylosaurs *Maleevus*, *Talarurus*, and *Shamosaurus*, and a strange dinosaur with huge, scythe-shaped front claws that was called *Therezinosaurus*. At the same time, their Mongolian colleagues Barsbold and Perle were describing other ostrich dinosaurs such as *Harpymimus* and *Garudimimus*, "egg-eating" oviraptorids such as *Ingenia*, and new forms with strange pelves, called *Segnosaurus* and *Enigmosaurus*, that have turned out to be related to *Therezinosaurus*.

After these great missions to the field, other fossil hunters took up the chase from 1963 to 1971. A series of expeditions to Mongolia was organized by the Institute of Paleozoology of the Polish Academy of Sciences. Their goal was to look for Mesozoic mammals in the Late Cretaceous beds of Bayn Dzak, under the leadership of Professor Roman Kozlowski. These expeditions took place in close cooperation with the Mongolian Academy of Sciences, and from 1964 and 1965 onward they were led by Dr. Zofia Kielan-Jaworowska, a woman of extraordinary energy and ability. Under her direction, with two other women scientists from Warsaw, Halska Osmólska and Teresa Maryańska, the geologist Gradzinski, and from the Mongolian side Barsbold and Dashzeveg, the Polish–Mongolian expeditions ended up with a most successful harvest of all kinds of fossils. In the course of three field seasons they collected an impressive range of dinosaurs and mammals, 35 tons in all, including a remarkable sample of mammals from Bayn Dzak, especially insectivores and multituberculates. They also collected a new sauropod, *Nemegto-saurus*, six more or less complete skeletons of *Tarbosaurus baatar* (*baatar* means "hero" in Mongolian), and the forelimbs of a giant ornithomim-id dinosaur, *Deinocheirus*, that were over eight feet long (2.5 m). Then, between 1967 and 1971, five expeditions brought back 60 skulls of tiny mammals, 200 lizards, an imposing array of carnivorous dinosaurs, os-trich dinosaurs, armored dinosaurs, and from the Tugrig locality, two skeletons apparently locked in mortal combat – the small carnivore *Ve-lociraptor mongoliensis* clutching the frilled skull of a *Protoceratops*.

After this series of expeditions, Kielan-Jaworowska came to spend a year in our lab in Paris, and we learned firsthand from her what the fieldwork in Mongolia had been like. We asked her along on our own expedition in search of the "Dinosaurs and Mammals of the Gobi Des-ert" so that we could benefit from her great experience on Mesozoic mammals. The prospect of crossing the desert sands and visiting the fa-mous Gobi sites, of which we had heard so much, absolutely thrilled us, and the excitement was tangible as our Franco-Italian expedition mem-bers gathered at Paris's Roissy Airport to board the plane for Moscow that would eventually take us to Outer Mongolia. On that day at the end of June 1991, Giancarlo Ligabue, the sponsor of our expedition and himself a trained paleontologist, Viviano Domenici, a journalist from Italy's famed newspaper *Corrière della Sera*, and Alberto Angela and Ser-gio Manzoni, who were making a film for television about the Gobi fos-sil beds, joined up with the French part of the crew, which included paleomammalogist Don Russell, paleoherpetologist Bernard Battail, and myself.

A stop at Moscow afforded us a brief time to see the capital of what was then the Soviet Union, especially the superb museum of the Acad-emy of Sciences, where the collections from the Soviet–Mongolian ex-peditions are kept. The atmosphere was very *fin de règne*, as if the immi-nent collapse of the Soviet Union were already felt by its populace. At the airport, our porters left their normal duties on the spot and were in-stantly transformed into chauffeurs of private taxis. In the dining rooms of the massive and Kafkaesque Hotel Russia, we were offered black-market caviar by the waiters, and at night a horde of young Russian girls offered themselves in the hotel corridors, under the eyes of the matrons who kept watch on each floor. On Red Square, Pierre Cardin was show-ing off his latest designs, and on our Aeroflot flight from Moscow to Ir-kutsk the vodka flowed like water – even in the cockpit.

After flying over Siberia and then Lake Baikal, we met up with Rinchen Barsbold, director of the Mongolian Geological Survey, and Khishigjavyin Tsogtbaatar, director of the paleontological wing of the Mongolian National Museum, to plan our trip of 5,000 km (over 3,000 mi.) into southern Mongolia. We were to make good use of the four Soviet-built all-terrain trucks offered to the Mongolian Academy of Sciences by their erstwhile big brothers. These were very strong and could go anywhere, but they had one drawback: The passenger seats were located over the engine. We soon learned, to our immense discomfort, just how hot, dry, and stressful this design was for the drivers and the crews. No doubt it was quite practical in winter, but in the Gobi desert, where summer temperatures reach 40 °C (104 °F), the truck's cabin, combined with the heat from the rumbling engine, proved an excellent weight-loss regime, complete with vibro-massage! But we were quite happy to be able to use these precious vehicles, thanks to Barsbold's influence, as well as the gasoline coupons that allowed us to buy fuel at the government service stations. Those were difficult times in Mongolia because when the Soviets pulled out, they took their economic aid with them.

After several days our crew set out in our well-stocked trucks on the trail southeast. We had been joined by four very nice Mongolian drivers whose names were very difficult for westerners to pronounce, and by a charming cook, Erdenechimeg Jambaldorji.

The immersion into the wilderness was absolute: We had only to bypass a few suburbs of Ulan Bator to find ourselves completely engulfed by the steppe. Shimmering heat waves arose as far as the horizon, shifting in the ever-changing wind. A symphony of greens played against the landscape in the magnificent high-altitude sunlight, and shadows of clouds accented the relief of every hillside. The meadows were filled with flowers: millions of edelweiss, chamomile, and wild rhubarb. Multitudes of various rodents, lemmings and marmots, dove into their holes as we approached. Kites soared overhead, while bustards cavorted with their young in the high grass. Everything enchanted and delighted me; I wanted to hike for hours and days in this wilderness, which was so diverse in spite of its apparent uniformity. *Baïarlalla!* Our Mongolian hosts had certainly outdone themselves in providing this incredible vista. That evening, I rolled out a rug of soft felt, decorated with geometric motifs, and slept on the ground under the stars. I thought of Jacques Bacot, who had explored the foothills of Tibet, and of his photos of those travels, which we'd just exhibited at the Museum of Man. In 1912, he wrote:

It seemed that I had always lived like this, and I wanted it to last forever. I wanted the unknown world to be without limits and, each day for countless years, for the dragons of my tent to rear up in the air of a new land. To travel like this is to live two lifetimes; to stop, to remain, is to live half-dead. In days gone by the voyages were long. Marco Polo's lasted twenty-seven years. Those were the days!

Outer Mongolia is three times as vast as France, but it only has 2.2 million inhabitants, of which nearly half a million live in the capital, Ulan Bator. So the country is nearly empty – but not completely, because everyone raises sheep, goats, yak, and horses: 12 million domesticated animals, of which 6 million are horses. The shadows of Genghis Khan's horsemen, who conquered a substantial part of the world, still flit across the Mongolian steppes. When you see these acrobatic horsemen, who at first seem so fierce to Western eyes, mounting their small, excited horses at a full gallop with no saddle, you can easily imagine the terror that our distant ancestors must have felt upon learning that the Mongols were getting ready to invade Europe again. And yet, on the steppe today, the hospitality of the nomad is equally legendary. The traveler passing by camp is invited to stop and enter the *ger*, a tent made of walls of camel skins covered on the outside by flapping white silks. Inside the *ger* he'll be offered *harack*, fermented mare's milk – delicious, sour, refreshing, and diuretic to boot – with a few flies floating on top for extra protein. Little cheeses, and eventually enough Mongolian vodka to knock someone over, complete the host's duties to his guest.

Seeing these nomads in their milieu, I could understand their preference for the steppe; it was difficult to see what advantages their less fortunate compatriots had, living in the horrible low-income housing units in the sad suburbs of Ulan Bator.

The first stop, after a jaunt of 500 km (about 300 mi.), was at the Early Cretaceous locality of Khuren Dukh. This place had already yielded remains of ornithomimid or ostrich dinosaurs, whose skeletons, in terms of overall bearing, evoke those of today's ratite birds: the ostrich, emu, cassowary, and their relatives. The first ornithomimids are found in the Early Cretaceous, recently evolved from their more primitive ancestors among the small, gracile carnivorous dinosaurs. Remains of fishes, turtles, and aquatic reptiles called "champsosaurs," which looked a lot like crocodiles but whose anatomy was very different, were also found here. Champsosaurs seem to have filled the same ecological roles as croco-

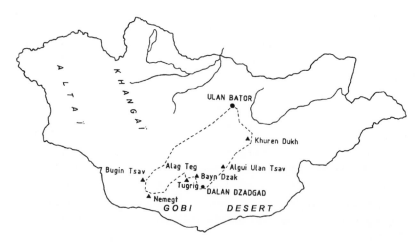

Figure 25. Map of Outer Mongolia, with the route taken by the French–Italian–Mongolian expedition in 1991.

diles, especially the long-snouted gavials, and they lived in Asia, Europe, and North America until well into the Paleocene – that is, after the Cretaceous extinctions and into the Tertiary Era. At Khuren Dukh the Mongolians had also found the remains of an ankylosaur and a hadrosaur. In the hollow of a little ravine, some bones appeared near the surface; a few minutes of digging and we took out a well-preserved arm of an *Iguanodon orientalis,* the Asiatic cousin of our European *Iguanodon* and our African *Ouranosaurus.*[2] All the parts of the arm were still connected: The hand with its metacarpals (palm bones), the fingers, and the claws were easily collected, and our Mongolian colleagues promised to come back and collect the other parts of the arm and – who knows? – even the rest of the skeleton, if it remained.

At the foot of the hill, not far from a nomad tent, a bunch of camels were rolling in the dust; they reminded me exactly of the dromedaries in Niger, though the latter's humps were smaller. Bactrian camels, African dromedaries, and the llamas of the Andes – a nice example of the evolutionary process unfolding on three different continents, resulting

---

[2] My British colleague David Norman has recently shown that *I. orientalis* is very likely the same as our familiar European *I. bernissartensis* – which adds an interesting new biogeographic slant to the problem.

Figure 26.   On the road, crossing the dune fields of the Gobi Desert. (Photo by P. Taquet)

in three distinct branches of the camel family, separated by continental drift.

Farther south, we set up camp not far from the Early Cretaceous locality of Algui Ulan Tsav, which was well known for its eggs and nests of sauropod dinosaurs. We were now at the edge of the Gobi. The word *gobi* is from the Mongolian language, Khalkha, and it means "basin." It's a generic term with subdivisions; so there can be red gobis, yellow gobis, and so on. The gobi that we were heading for is a semidesert, mostly of steppe; the south and west are more desertlike, but the north, where we were, was still steppelike.

Early in the morning, we set out into an immense labyrinth of hills that were deeply cut by erosion. The heat was torrid. Our Mongolian colleagues taught us a trick for finding a nest of sauropod eggs; their method was original, but effective. Wandering the bottoms of each little ravine, eyes fixed on the ground, you could with a little luck find large spheres of rock the size of bowling balls. These were eggs that erosion had released from the loose matrix; they harden as they weather out, drying in contact with the air, and they roll easily down the slope.

Just by retracing the egg's path, you can usually find where it came from – and the rest of the nest still in place! Don Russell, with one of the drivers, put the method into practice, and soon shouted over to us. He had found the goose with the golden eggs – or at least a nest with six sauropod eggs still in it! We were lucky, and delighted. The eggs, about 20 cm (8 in.) in diameter, were very well preserved, their shells decorated with a light coating of tiny ornamentations. These little ornamentations were useful: They prevented sand grains from blocking the pores that allowed the egg to exchange gases with the outside air, so that it could develop and produce a normal embryo.

The first discovery of unquestionable dinosaur eggs had taken place in Mongolia, at the foot of the cliffs at Bayn Dzak, exactly where we were now heading. At the foot of this cliff, the Andrews and Granger American expedition had discovered entire nests of about 20 oblong eggs grouped in pairs. Near these nests they had found great numbers of complete skeletons of *Protoceratops*. These animals were around 2.5 meters long (8 ft), with a triangular skull and a toothless beak. In front, above the beak, was a rostral bone with a rough surface that might have supported a horn of some kind when the animal was alive. At the back of the skull, a large bony collar might have played a role in protection or display, or it might have anchored powerful muscles for chewing tough plant matter.

Protoceratops andrewsi was named in 1923; it is considered close to the base of the lineage that culminated in *Triceratops* and the other famous North American ceratopsian dinosaurs. The Americans also thought that it was responsible for the eggs found in such great quantities at Bayn Dzak, a discovery that created an enormous public stir at the time. Ironically, new expeditions to the Gobi Desert, also from the American Museum in New York, including Mark Norell, Mike Novacek, Jim Clark, Luis Chiappe, Malcolm McKenna, and our Mongolian colleagues Perle Altangerel and Dashzeveg, have recently turned up nests and eggs of this same type – with an *Oviraptor* brooding on top of them! It appears that the legendary "egg-stealer" – the name Osborn gave to *Oviraptor* – was actually the protective parent. However, the source of the conflict between the "fighting dinosaurs," *Protoceratops* and *Velociraptor,* can still only be guessed at.

As it happens, the first dinosaur eggs were found in France, long before the Mongolian expeditions; but at the time, people could only won-

Figure 27. Sauropod dinosaur eggs at the Early Cretaceous Algui Ulan Tsav locality in the Gobi Desert, Outer Mongolia. (Photo by P. Taquet)

der what sorts of incredible animals might have laid them. It was the south of France, in Provence and Languedoc-Roussillon, that was graced with the first discoveries of evidence of the results of the mating rituals of dinosaurs. The eggs were first found in the Roussillon in 1859 by the Abbé Pouech, a local priest. Then in 1869 the geologist Philippe Matheron described well-preserved eggs collected near Marseilles. He supposed that a giant bird or some kind of great reptile must have laid these eggs, which were 20 cm in diameter (8 in.). The eggs were eventually attributed to a sauropod dinosaur, a titanosaurid named *Hypselosaurus priscus*. The eggs from Algui Ulan Tsav, which are not the same age as those from Provence but from the Early Cretaceous, nonetheless resemble those of *Hypselosaurus priscus* in their general shape and size. It's a real problem to assign eggs to a specific dinosaur, because unless you're lucky enough to find a fossilized female in the act of laying her eggs – which so far hasn't happened, although the new *Oviraptor* discovery is about as much as we could hope for – it's impossible to be sure of who the egg layers were. Eggshell structures are quite different, not just among the dinosaurs but among living animals. A bird egg doesn't have the same structures as a crocodile egg. In the Gobi Desert there are some 15 kinds of eggs that vary in size and shape. Some are from crocodiles, others from lizards, turtles, and probably birds, but the dinosaur eggs show a lot of variation in structure. This probably reflects species differences, but there are also changes in form and crystallization during and after fossilization, so the situation is not straightforward.

At Algui Ulan Tsav, the dinosaurs were laying up to 20 eggs at once, arranged on the ground in three circular rows. Did they lay them in the open air in a little hollow of sand, like ostriches do in the desert today? Did they dig holes and bury the eggs, like turtles and some crocodiles do? Or did they cover them with rotting plant material that would give off heat and help incubation, like some birds do today? One answer to this came from the Australian zoologist Seymour, who showed in 1980 that the diameter and density of the pores on eggshells can tell us some interesting things. In fact, there's a correlation between the transport of gases through the shell, the loss of water during incubation, and the diameter of the pores in bird and other reptile eggs. Seymour calculated the rate of water-vapor transport and of gas transport through dinosaur eggshells. When you compare them to the eggs of today's birds and other reptiles in which the nesting conditions are known, it turns out that three types of dinosaur eggs are very porous, suggesting conditions of high humidity, low oxygen, and high carbon dioxide. (Such eggs probably would have dried out in desertlike environments.) Seymour found that these conditions were found in eggs that developed underground or in nests that were covered with material that permitted incubation.

But now Jack Horner's remarkable discoveries in the Late Cretaceous deposits of Montana in the United States have helped us make enormous progress in our knowledge of the earliest stages in the lives of dinosaurs. Horner and his colleague Bob Makela were lucky enough to get their hands on several areas where dinosaur nests, eggs, and hatchlings are found. These nests were covered by an eruption of fine ash and cinders that preserved the bones very well. The researchers were able to study eggs, nests, embryos in the eggs, hatchlings, juveniles, and adults both in the field and in the laboratory. Some of the finds came from a hadrosaur that they called *Maiasaura,* and some were of a hypsilophodontid – a small, bipedal, herbivorous ornithopod with both slicing and grinding teeth – that they named *Orodromeus.* (Some of the eggs and embryos ascribed to *Orodromeus* they later reattributed to a small carnivorous dinosaur, *Troödon* – in fact, their original suspect.) Horner and Makela's work showed that there were three types of eggs in the Montana deposits: The first two types were ovoid, planted upright in the sediments, and only partly covered; the third type was rounder, laid in two rows, and entirely covered with sediment. It appears that certain species, like *Maiasaura* among the hadrosaurs, *Orodromeus* among the hypsilophodontids, or *Protoceratops* among the horned dinosaurs, laid their eggs in flat places right on the ground, whereas others, like the sauropods, laid their eggs in rows and buried them.

In the case of the so-called *Protoceratops* eggs in Mongolia, which now are known to belong to the theropod *Oviraptor,* we find the eggs set out in pairs, radially, forming a series of circles. Some of these circles contain 10 or 12 pairs of eggs. Rinchen Barsbold thinks that the females followed a circular trajectory as they laid the eggs, and because their two ovaries had the same developmental timing, they laid two eggs at a time. This differs from the situation in living birds and other reptiles, where one of the ovaries is reduced and nonfunctional.

The red hills of Algui Ulan Tsav formed a backdrop worthy of the best Western movies. But of course you'd have to call them "Easterns" here, and so our Polish colleagues did, when they made a little film about their previous work in the Gobi Desert. But now we got back on the road that led southeast to the city of Dalandzadgad. After filling up our four trucks, we headed west this time, to make camp at the celebrated Flaming Cliffs of Bayn Dzak. We arrived just at the end of the day, when the superb cliff was inflamed like the scenery from a grand opera under a spectrum of red, pink, and orange light.

But as we approached, another caravan of all-terrain vehicles came into view. Now, when two caravans cross in the desert, it's natural for the travelers to stop and have a quick chat. But we were astonished to see that the travelers getting out of their vehicles were not Mongolians at all, but Americans! Yes, Americans, wearing cowboy hats, jeans, and checked shirts. We had run into our colleagues from the American Museum of Natural History in New York, who had already spent two seasons following the trails of their illustrious predecessors in the Gobi Desert from decades ago, along with their Mongolian colleagues whose more recent work had been so successsful. Almost 70 years after the first trip to the foot of the Flaming Cliffs at Bayn Dzak, here they were, taking advantage of the recent political thaw to follow in the footsteps of Andrews and Granger. One of their expedition leaders was the great specialist in Mesozoic mammals, Malcolm McKenna, whose financial connections allowed them to fund extensive field campaigns like this one. There was also Jim Clark, a specialist in fossil reptiles whom I knew well, Mark Norell, who had worked with French crews in the Mesozoic faunas of Cameroon, and the Mongolian paleontologist Perle Altangerel, who has an intimate knowledge of the Gobi dinosaurs. The world of paleontologists is indeed small, but our meeting was unusually special, particularly when the American Malcolm McKenna and the Frenchman Don Russell spotted each other and laughed and hugged joyfully. The reason was simple: Don, who originally came from North America, and Malcolm had both been trained at the University of California at Berkeley. They had studied paleontology together and were old friends, even though their careers had taken them to opposite sides of the Atlantic. We all greeted each other and took photos immortalizing the brief encounter, then piled back in our trucks and headed back through the waves of Gobi sand.

We planted our tents at the foot of Bayn Dzak. The surroundings were even more grandiose than at Algui Ulan Tsav; not far away, John Ford had filmed *Lost Patrol* (1934). I took out my copies of the black-and-white photos taken by the American Museum expedition in 1923, and it was striking to see that the cliff face was practically unchanged after 70 years. You could recognize each sandstone turret, each crag, each jagged outline along the plateau. Rainfall in the region is rare, virtually nonexistent. During the evening, the skies darkened and huge black clouds descended on us; we could see the rain falling in sheets, but the dryness and the evaporation from the ground were so strong that the rain never reached us. It disappeared by and by as it fell. As if to vent the heavens' frustration, a brief but violent tornado of wind suddenly blew through the camp, bearing away tents, tables, dishes, and papers; one of us was grazed by a falling pole.

The next day, our prospecting brought us among the slopes and fallen boulders at the foot of the cliff. This is where the Polish expeditions had found the remains of tiny Mesozoic mammals. These mammals were very rare, for two reasons: They were relatively few in number, and more important, they were very small – most were mouse-sized, and almost none grew larger than a cat. Their skulls were hardly ever more than a few centimeters in length, and their teeth were only millimeters long. This is why looking for them requires so much patience. One method is to collect sediment, screen-wash it, then pick through the resistant debris for tiny pieces of bone and teeth, using a binocular microscope. Out in the field, you can try to find more complete pieces by stretching out on the ground and scrutinizing each block of matrix, each corner of rock disengaged by erosion, scanning the surface for the traces of small bones still trapped in the rock. In the first case, given that the beds are relatively rich in fossils, you might find a tooth or two each week, and in the second case, you might find a skull each day, or maybe each week. But apparently the gods of paleontology were with us, because in only a single afternoon Don Russell and I found a complete skull of a little mammal. After hours of crawling along or creeping on hands and knees, suddenly to recognize a small fragment of globular sandstone 2 cm across, holding your breath, picking it up delicately to inspect it under a hand lens, finding that you can distinguish the tiny rows of teeth with their enamel crowns gleaming in the sun, picking out a few suture lines between the skull bones . . . to find after all what you have come so far to find, evokes a beautiful and incomparable feeling. It's like what a botanist feels when he finds the orchid of his dreams, what the ornithologist feels when the Great Auk alights at her feet; it's what all Nature lovers hope to discover whenever they allow themselves to be seduced by her marvels.

Most people don't realize that mammals lived at the same time as the dinosaurs, and that both groups started to populate Earth at about the same time, which was at the end of the Triassic and the beginning of the Jurassic, about 200 million years ago. Mammals and reptiles have very separate lineages, reaching back to when vertebrates first came out on land over 300 million years ago. The two groups are very different in the construction of their skeletons, their physiologies, and even in the development of their brains. Because mammal brains develop so quickly, two bones that were in the jaws of the first tetrapods (and still remain there in other tetrapods today) have migrated into the ears of mammals, to become part of the sound-conducting apparatus. The teeth of mammals provide further evidence of difference: In reptiles, the teeth are generally all pointed and conical; they seize and grip prey, manipulating it in the mouth before it can be swallowed whole. In mammals, the teeth are differentiated into incisors, canines, premolars, and molars. The very first mammals, or things that were very close to mammals, gave rise to three distinct groups of true mammals. The first had molars with three *tubercles,* or cusplike crowns, running front to back on the tooth: a small one followed by a large one and then by another small one. These mammals are generally called "eotherians" (from the Greek *eos,* or "dawn," and *therion,* "beast"). The second group had molars with two or three parallel rows of tubercles running from front to back; these were called "allotherians" (from the Greek *allo,* meaning "different" or "other"). The third group had molars with the tubercles arranged in the shape of a triangle; they were called Therians, and they include us and our distant ancestors, as well as most mammals living today.

The "allotherians" enjoyed a brief success through the course of the Mesozoic up to the end of the Jurassic, some 140 million years ago. One group that arose within their ranks, the multituberculates, is known almost as well by its nickname of the "rodents of the Mesozoic." They had sharp cutting incisors and grinding molars, much like those of living rodents. More multituberculates are found in the Gobi Desert, in the 70-million-year-old beds of the Late Cretaceous where we were now working, than anyplace else on Earth. The two skulls that we had just collected belonged to this group, one to the genus *Sloanbaatar* and the other to the genus *Kryptobaatar.* It was the first time that I had found the complete skull of such a small animal, let alone a Mesozoic mammal; after 30 years, I was mostly used to dinosaur bones, which grew to more respectable dimensions. The discovery of these two skulls, and the conditions of their fossilization, allowed us to take a great number of dramatic and instructive photographs for our exhibits in Venice and Paris, which we could juxtapose face to face: Tiny mammals with 2-cm skulls against giant dinosaurs like *Tarbosaurus,* whose skull is more than a meter long.

This relay race between the mammals and the dinosaurs lasted the entire Mesozoic Era. It seems that the small size of the mammals, their mostly nocturnal mode of life, their agility, their intelligence, and their ability to find food allowed them to diversify little by little, and eventually to conquer the entire planet. The mammals bided their time – which was after the dinosaurs disappeared at the end of the Mesozoic Era – to occupy many of the adaptive zones of their predecessors and to give rise to a variety of forms of all sizes and shapes, from elephants to whales to bats, in the seas and in the air as well as on land – as dinosaurs had never done (apart from their living members, the birds).

One of the keys to their success was the invention of the placenta. This organ ties the embryo to the uterus during development, assures it a good supply of oxygen and food, and allows it to stay warm and safe inside the mother for as long as possible. Marsupials don't have a placenta, so in the kangaroo and the koala and the opossum, the development of the young ends by dumping them into a pocket outside the mother's stomach, the marsupial pouch. Many other characters, such as the absence of udders, and differences in the skull, the teeth, and the skeleton, distinguish marsupial and placental mammals from each other.

The oldest known placental mammal is *Prokennalestes* (whose name is a tribute to Malcolm McKenna, whom we have already met out in the field in the Gobi). It was found in the Early Cretaceous beds of Mongolia, a small animal whose skull was only 3 cm long. This animal was very close to the ancestry of all the other placental mammals, which are known from many forms in Central Asia. Very rapidly – on the geological scale, of course – two lineages diverged at the beginning of the Late Cretaceous. One was the insectivores, which eventually branched into several living groups of mammals, including moles, shrews, and hedgehogs, as well as primates; one of their first members was called *Kennalestes* (also in honor of McKenna). The second lineage is generally known as the condylarths, which have traditionally included the forerunners of all the hoofed animals, including the elephants, rhinos, horses, ruminants, pigs, and camels, as well as a host of extinct forms. In the course of long and difficult research on the first "modern" mammals, with their triangularly arranged molar tubercles, our paleontology laboratory at the Paris Museum had earned its place in the first rank of international scholars, thanks to the labors of Denise Sigogneau-Russell and her husband Don Russell in the Early Cretaceous beds of the Moroccan High Atlas. With the help of Michel Monbaron and his team of Moroccan technicians, they collected a new and extremely interesting and important Mesozoic mammal fauna. It included the most ancient *tribosphenic* mammalian molar – that is, a mammalian tooth that cuts and grinds at the same time – with the development of a *talon,* or heel, on the lower molars and an extra bump, or *protocone,* on the upper molars, which accomplished the grinding function.

In the end, the great and powerful dinosaurs surrendered Earth to the mammals, who were smaller and more modest but more intelligent. Much of this drama was played out on the Late Cretaceous, now semi-desert plateaus of Mongolia, and the cast of characters included our most remote ancestors among the mammals.

Leaving this superb and grand site with regrets, we picked up the road again westward as far as the Tugrig plateau, whose beds also date to the Late Cretaceous. This locality is also rightfully celebrated for the discovery of the "fighting dinosaurs," the fossilized combat between *Velociraptor* and *Protoceratops.* The predator died with its forelimbs clutching the frill of the skull of its apparent prey, while its hindlimbs were trying to rip open the animal's chest and abdomen, using the enormous claw that it bore on the second toe of each foot. For some unknown reason, the two adversaries died in combat, and their bodies sank rapidly into the sandy deposits of a lake. The Polish–Mongolian expedition discovered this fossilized combat, or paleodrama, 80 million years later, and the striking tableau is now one of the most spectacular pieces in the collections of the Museum at Ulan Bator – or indeed, in those of any museum.

Here at Tugrig the sands were white, blinding under the sun; the plateau where we camped was battered by the violent winds of the Gobi. It was cut by an immense cliff, and early in the morning our crew members set out, each in a different direction, to comb the shallows, ravines, and promontories in search of fossils. My own path led me to a steep slope, where I saw a few bony fragments barely sticking out of the ground. What might this be? Drawing closer, I hunched down on a stretch of sand that looked like a nice seaside beach, although here in the Gobi the last tide had gone out more than 200 million years ago. Very carefully, I separated some grains of sediment from the bone fragments; they were pieces of a smooth bony surface. Using my brush, I could reveal more of the bone surface without disturbing the flat plate of bone, which was becoming longer and more flared. Within a few minutes I had exposed a well-preserved surface of a few dozen square centimeters. Using my knife, I disengaged the sediment with care. This was fairly easy because only a few centimeters of sand covered the bone, and I saw that this mysterious fossil was more complete than I had expected. I realized all of a sudden, as I was clearing off a scalloped edge of bone, that I had just discovered part of a skull. Holding my breath, I continued to dig almost feverishly. A large cavity that had once held the eye appeared little by little; then I could follow the upper border of the skull, and a rounded bulge indicated the upper end of the snout. The border curved strongly downward, and at once there appeared the outline of a snout pierced by a long nasal opening. This was incredible! The sand was so soft that I could brush it away from the skull with my hand. In short order, the right cheek, the right maxilla, the row of maxillary teeth, the right mandible complete with its teeth and articulating with the upper jaw, then the whole right side of the skull of a superb *Protoceratops* appeared on the hillside. Absolutely complete, magnificently preserved, this animal had laid in the sand for 72 million years. And now here it was, looking at me, its eye turned in my direction, almost smiling at me with its clenched jaws, as if to say "Hello, here I am!" I was stupefied. Staggered. It had taken me only an hour to expose this marvel, this specimen that was so representative of the quality of preservation that you find in the Gobi Desert fauna. I called my companions over so they could admire this splendid dinosaur skull, and so that Sergio Manzoni could film a nice sequence showing how the exceptional conditions of Tugrig allowed this kind of preservation.

This locality had yielded dozens of complete skeletons of *Protoceratops,* in which all the bones stayed in articulation. But there was something else unusual about the Tugrig locality. These skeletons were all in a vertical position, as if the animals had remained standing on their four feet. The *Protoceratops* lived on the shores of lakes, and all these individuals had been entombed in the swampy borders of those lakes. Two specimens in the museum at Ulan Bator are particularly spectacular: Prepared with great skill by the Mongolian technicians, they are exhibited in their death positions, and only some sandstone supports hold the skeletons in place. They weren't missing even a toenail: Even the two little clavicles, the shoulder bones whose very existence in dinosaurs was long in doubt, were in their correct anatomical place. (The presence of a clavicle in dinosaurs is of interest because it bears on the argument that establishes the ancestry of birds among dinosaurs.)

Protoceratops andrewsi is certainly one of the best-known dinosaurs in the world. Since the 1920s, the American Museum expeditions collected many specimens of this species, and they assembled a growth series from newborn to adult that has been displayed ever since in their exhibit hall. The abundance of *Protoceratops* specimens permitted the study of intraspecific variation, and 40 different variables in a biometric study showed how the proportions of the skull change during growth, and how to tell the difference between the sexes. The Tugrig locality also yielded many skeletons of baby hadrosaurs, including newborns 30 cm long (1 ft.); one of them, a superb specimen, would be exhibited in Paris.

After our visit to Tugrig, which had been so exhilarating for all of us, our tour continued, always westward. We crossed mountain chains, rocky passes, and strings of sand dunes. Here and there, at the edge of the trail, where the crossings were most difficult, our drivers would stop their trucks in front of a pile of rocks; they would get down and walk around it three times, clockwise, tossing a few coins or matches as an offering. This pile of rocks was an altar, an *obo*, consecrated to the spirits whom one asks for protection. Our Italian friends suggested that their word *obole* has its origin in the Mongolian word *obo*, which Marco Polo brought back to Europe, along with the recipe for pasta. I don't know if this is true, but in any case we can testify that the Mongols love their pastas, and they are very talented in whipping them up, even in the middle of the steppe. And these are real pastas, from which they prepare big raviolis stuffed with chopped mutton; real pastas, flattened and cut to perfection into large strips, using a machine that you see all the time in Italian restaurants – but hardly something that we expected to find in the steppes of Central Asia.

We finally reached the famous Nemegt Valley, where the Russian and then the Polish–Mongolian expeditions had collected so many fossils. Erosion weathered out new fossils every day, and the whole region was renowned for its richness in dinosaur skeletons. This is where they found so many ornithomimids and tarbosaurs. At the foot of our camp, a deep canyon stretched away; at the bottom of this canyon, the next day, our Mongolian colleagues brought us to examine the back half of a superb skeleton of one of the giant carnivorous dinosaurs. All the tail vertebrae, all the hip bones, and a good part of the hindlimbs were embedded in the sandstones in perfect articulation.

The tarbosaur, *Tarbosaurus,* is one of the largest carnivorous dinosaurs ever known to exist. It was part of the tyrannosaurid lineage, a close cousin of the American *Tyrannosaurus.* In fact, the two genera are very similar, and if they hadn't been described by different paleontologists on different continents at a time when politics made it difficult to exchange information, the second genus would probably have never been created. The fact that the specimens of one genus are in Moscow and Ulan Bator, while those of the other genus are in North American museums, has still frustrated direct comparisons. *Tarbosaurus* has two tiny front limbs with two fingers, the first with two phalanges, the second with three. All of the animal's power is concentrated in its massive jaws and its powerful hindlimbs. The high skull is mainly distinguished by its huge openings: the large nares (for the nostrils) in front, then the big antorbital *fenestra,* or keyhole-shaped orbit that housed the eye, and finally the two temporal openings. The neck is stockily built, and its vertebrae have strong apophyses. The three-pronged pelvis has a stout pubis, whose distal end is provided with an enlarged triangular "foot" or "boot." In the hind foot, the three metatarsals, or sole bones, are raised off the ground as in all dinosaurs, and are pinched and interlocked as in the ornithomimids, their close relatives, to reinforce the strength of the foot. This carnivorous dinosaur could reach 10–12 meters in length (33–39 ft.). It was described for the first time by the Russian paleontologist Maleev. Nearly 30 specimens of this carnivore have already been collected, and its anatomy is well known. Rinchen Barsbold showed us something unusual that he had discovered about this specimen, however: The impression of the reptile's skin on the sandstone had been preserved, thanks to the exceptional conditions of fossilization. Here, when the animal died, it was buried such that its skin still clung to its bones, no doubt stretched and dried out like the camel carcasses that you can see in the desert today. The dry climatic conditions, added to the fine-grained sediment that faithfully preserved every detail, provided us with a pretty good idea about this dinosaur's skin. There were already well-known impressions of fragments of skin from herbivorous dinosaurs, including European iguanodontids and American hadrosaurs, where complete "mummies" had been collected. The Ulan Bator museum also had some beautiful pieces with hadrosaur skin, but this was the first time to my knowledge that a carnivorous dinosaur's skin was preserved. It seemed to be made of small scales, whereas the ornithopod dinosaurs' skin comprised large scales, with islands made of larger platelike scales.

The vision of this tarbosaur, this "terrible reptile," as its name denotes, encased in the sediments of this canyon, reminded me of the text of a science-fiction novel written by our colleague Ivan Efremov in 1953, *The Shadow of the Past,* inspired by his research right here in the Gobi Desert:

This burning valley, black, lifeless, strewn with giant bones, had something strange about it. It called up the old legends that spoke of combats with dragons, the tombs of titans, monsters buried together by the flood. And one learned soon enough the meaning of these stories, which had certainly taken their origins from a similar discovery. . . . Nikitine guided Miriam toward the excavation, and the astonished girl beheld the sprawled skeleton of an enormous reptile.

In this story, Efremov imagines that the rock face of a cave in the mountain, coated with resin, has served as a photographic film on which the features of an enormous carnivorous dinosaur have been permanently fixed, because the cave has become a natural darkroom. Nikitine and Miriam, the Russian paleontologists on the expedition, watch in fascination as the rays of the setting sun bring out the apparition of this monstrous dinosaur . . . first in bone, and then in the flesh! The dream of every dinosaur paleontologist.

For us today, in the twilight of the end of a beautiful day, in the hollow of this canyon in the Nemegt Valley, in front of the parched remains of this superb tarbosaur with its skin still preserved, truth became stranger than fiction, and the shadow of the past was truly present.

A BONE HUNTER
IN LAOS

N 1927, a French geologist, Josué-Heilmann Hoffet, set out for Indochina. He was attached to the Geological Survey at Hanoi, and after some initial work on the Kuang Chon Wan Peninsula, which had been surrendered by China to France in 1898, he wound up in charge of mining surveys of the continental formations of lower Laos, the region east of the city of Savannakhet. Hoffet was a young and brilliant researcher of 26; he had taken the geology course taught by Paul Fallot at the University of Nancy. He was a field man, and for 22 months he set out in all directions over an immense territory covered with forests – on foot, on horseback, even on an elephant. At the time, the region was still not secure, and the French administrators that dispatched him to this remote outpost were not about to take any responsibility for his well-being. A telegram addressed to him at Attapu, in southern Laos, by the General Director of Mines at Hanoi, simply said:

Approved but understood geologist Hoffet proceeds at his own risk.

In lower Laos, Hoffet marshaled a great store of scientific information, and he submitted his thesis to the Sorbonne. It was published in 1933 as a memoir of the Geological Survey of Indochina entitled "A Geologic Study of the Center of Indochina between Tourane and the Mekong (Central Annam and Lower Laos)." In 1935 two geologic maps (1: 500,000) of Hue and Tourane (today called Da Nang) were published; Hoffet had overseen their production. These field campaigns and labors completed, he was summoned east of the village of Muong Phalane, near Sepone, at the edge of Highway 9, which connected Savannakhet to Tourane; oil had been discovered there. So Hoffet was brought back to this area, and in the course of a new four-month field campaign, not far from Phalane, he collected some fossil shells of freshwater mussels, as well as some sparse fossil bone remains, including a caudal vertebra from what he called a "Saurian." Hoffet described for us the beginning of what was to become one of the strangest stories in the annals of paleontology:

Alerted by this discovery, the natives, who had often seen similar bones but had thought they came from buffaloes, told me about places where they thought such remains were to be found. There were a great many dead ends in this open forest, under the scorching sun of lower Laos. Sometimes the only thing to see was a stone that vaguely resembled a bone. But sometimes the signs panned out. I found a saurischian dinosaur pelvis here, a vertebra or a fragment of hadrosaur femur there. I searched in vain for miles around; I combed the trackless brush in all directions, but I never saw two associated pieces of bone within a radius of 40 km [25 mi.]. Finally after two weeks of work, I brought back about 100 kg of bone [220 lb.], mostly silicified, belonging to three different animals. Returning two years later, I was fortunate enough to find a complete titanosaur femur, but unfortunately it was destroyed by a tree that had taken root right in the middle of it. Four kilometers away, the head of a femur. Then a sensational find: 50 km [30 mi.] south of Muong Phalane, a collection of bones that seemed promising, judging from the natives' description. The next day I set out. At the last village before the site, they refused to take me any farther. The bones belonged to genies, and evil would befall anyone who removed them. After a whole day of discussion, we settled on a price that depended on the value of the fossils, with the addition of a buffalo to atone for my sacrilege. Early the next morning we left the village. I thought I would find the remains of an entire animal. You can imagine my disappointment after four hours walking, to find that my bones were only limestones sculpted into bizarre shapes by wind and streams. The journey was not entirely fruitless, though: on the way back I found a rocky hillock in which vertebrae, shoulder girdles, and other pieces were preserved. They were relatively common there, since the bones were only a couple of meters away from each other.

The discovery of the freshwater shells and dinosaur bones gave Hoffet what he needed to date the continental formations of lower Laos, which were called the "Indosinias," to a time later than the Paleozoic Era. Hoffet announced the discovery of Cretaceous-aged sediments in Indochina in a note to the French Academy of Sciences in 1936; ironically, it made parts of the geologic map that he had just published out of date. For this reason, Hoffet felt that he had to write a supplement

Figure 28. Excavating sauropod bones at Tang Vay, Laos, in 1993. (Photo by J.-F. Hoffet)

(dated 1937, published in 1938) to his 1933 memoir. He explained that in his earliest works he had concentrated especially on the mountain ranges and had neglected the flat sandy region, because his necessarily hurried reconnaissance of this vast area had suggested that it was fairly monotonous and held little of interest. Based on its general similarity and its facies, Hoffet had thought that it was of Liassic (Early Jurassic) age, from earlier in the Mesozoic.

It's worth noting that Hoffet had some valid excuses for his initial interpretation. Between the Mekong to the west and the Annamite Mountains to the east stretched a vast basin with very little relief, and outcrops were rare. The whole region was infertile, and its surface was covered with splayed alluvium. The drought from November to April al-

149

lowed almost nothing to grow except a forest composed almost entirely of dry-tolerant trees like *Dipterocarpus,* a resinous plant with tough, deciduous leaves, while the ground was covered with tall grass. Moreover, the waterways were bordered by thick forests of royal bamboo more than 15 meters high (50 ft.). This made it difficult to interpret the geology. In 1930 there were only eight people per square kilometer (three per square mile). Paths were few, and topographic maps were nonexistent.

During a trip to France, Hoffet settled in Strasbourg, near the Institute of Geology, and studied the clam shells that he had brought back from Laos. He also boned up on dinosaur anatomy, benefiting from the experience of the geologist Schneegans. He worked at Nancy with Fallot, paid a visit to Paris to see Professors Jacob, Piveteau, and Lacroix, and at Tübingen, Germany, he met the paleontologist Hennig, who had worked from 1908 to 1912 with the German teams who excavated the celebrated Tendaguru beds in Tanganyika, where a great many dinosaurs had been found, including the enormous sauropod *Brachiosaurus,* whose skeleton became one of the great trophies of the Humboldt Museum in Berlin. After this European stint Hoffet returned to Indochina, and he left his post in the Geological Survey to teach geology at the *École supérieure* in Hanoi. He was then able to publish the results of his research on the fossils that he had collected in Laos. He had put some distance between himself and his colleagues at the Geological Survey, as a result of a disagreement in 1941 with Jacques Fromaget, head of the Survey. Hoffet thought that the molluscs from Muong Phalane indicated brackish deposits of Campanian age (Late Cretaceous), whereas Fromaget thought that they indicated freshwater deposits that were earlier than the Late Cretaceous. It is perhaps not beyond suspicion that Fromaget was jealous of the compliments paid to Hoffet when his geologic map of lower Laos appeared. But in the rarefied atmosphere of the small French colony within Hanoi, and particularly within the Geological Survey, the presence of strong personalities did not help matters. There was a history of conflicts: Several years before, the memorable "Deprat affair," named for a member who had been unjustly accused by his superior of having invented his scientific results wholesale, had made a great stir, shaking up the geological communities of Indochina and even France.

Hoffet had to wait two years for the intervention of the president of the office of the Scientific Research Council, Monsieur Drouin, in order to secure the return from the General Inspector of Mines and Industry of the material that he had discovered, and to see his research through. In 1942 he published his "Description of Some Bones of Titanosaurians from the Senonian of Lower Laos," in which he erected a new species of dinosaur, *Titanosaurus falloti*, based on a complete femur collected from Muong Phalane. The titanosaurs or "titan reptiles" were Cretaceous sauropods whose first representatives had been described in India. Hoffet dedicated this new species to his teacher, Paul Fallot. Then he briefly described the most characteristic bones that belonged to ornithischian dinosaurs that he had found. (Like many workers of the time, he called ornithischians "avipelvians" and saurischians "sauripelvians"; these are equally descriptive names, but they have fallen out of use in recent decades.) Among the pieces that Hoffet had collected were several ilia, and the ilium is one of the very characteristic bones of the pelvis in dinosaurs. He had also collected vertebrae from the neck, back, and tail, and some distal ends of thigh bones, as well as remains of turtles, crocodiles, and fishes. He erected another new species of dinosaur, a hadrosaur or duckbilled dinosaur, that he had compared to *Mandschurosaurus amurensis*, an animal collected by the Russian paleontologist Riabinin, based on material collected from the banks of the Amour River in Manchuria. He also compared his new find to *Mandschurosaurus mongoliensis*, which had been described in 1933 by Gilmore, based on material collected from Iren Dabasu in Mongolia by the American Museum expeditions. Hoffet called his new species *Mandschurosaurus laoensis* to commemorate its Laotian origin. The features of this second dinosaur from Laos were what really convinced Hoffet of the Senonian (Late Cretaceous) age of these lower Laotian beds; such hadrosaurs were not known from earlier deposits.

Unfortunately the Second World War brutally ended Hoffet's brilliant career. Japanese troops occupied Indochina; the French Resistance got organized, and in March 1945 he joined the French Underground, which was trying to hook up with the Allied troops. Hoffet died during a clash with Japanese forces at the pass at Nui To, northwest of Hanoi. His wife and his four children waited in vain for his return, and they had no choice but to return to France after the war.

A silence of 45 years then settled on Hoffet's work. His publications were not widely circulated because they had appeared in a journal printed in Hanoi at the beginning of the war years, so they were known only by a handful of specialists.

151

As it happens, I had copies of Hoffet's two articles in my collection of the dinosaur literature of the world; I had bought them from a bookstore specializing in natural history that was selling off the scientific library of a retired colleague. Just looking at Hoffet's description of what he called *Mandschurosaurus laoensis* showed how important this material was. I was very familiar with the ornithischian skeleton, as a result of my work on *Ouranosaurus*, and the hip bones that Hoffet figured suggested to me that their characteristics were not those of hadrosaurs, but of iguanodontids. In 1940, these differences were not well understood, and it was not surprising that Hoffet, writing from his remote outpost in Laos, would have trouble distinguishing between the two. Nevertheless, his discoveries deserved to be reexamined, and I thought it would be tremendously important to revisit the sites that he had explored. We had asked the CNRS (National Center for Scientific Research) in vain to allow us to recruit a young worker to take up the Laotian research, even though we presented them with a most solid scientific dossier. Finally, I decided to go to Laos myself. I left in November 1990, freed from my administrative duties directing the National Museum of Natural History, after putting the renovation of the Gallery of Zoology on track by securing the funding to turn it into a modern museum that would center on evolution and conservation of the plant and animal world.[1]

Laos had suffered the chaos and destruction of war on the Indochinese Peninsula almost without relief since 1945. First there had been the Indochina War, then the Vietnam War; finally, civil war in Laos itself, culminating in the seizure of power by the Pathet Lao. The region where Hoffet found his dinosaur bones was at the western edge of the

---

[1] I should recount briefly some details of this story, which shows what can happen when paleontology meets politics. President Mitterand was attending a reception adjacent to the ancient Gallery of Zoology, a triumph of the Museum since the days of Cuvier and Gaudry, but which had been in disrepair and closed for nearly 35 years. Before leaving, Philippe asked if he could just have five minutes of the president's time, and this secured, he ushered Mitterand and his retinue through a door into the dark and sepulchral exhibit. Suddenly a bank of lights blazed, illuminating the gloom and decrepitude of what had once been a magnificent display of the diversity of animal life. "But this is terrible," Mitterand said. "It must be restored." Philippe of course agreed, and when asked, quoted the president the price estimated by their scientific and artistic consultants: 400 million francs (about $65 million). "When would it be ready?" was the president's next and most natural question. And Philippe had a ready answer: "Before the next election." The funding sailed through administrative channels. And if you have not yet seen the new Gallery of Evolution, it is stunning.

Ho Chi Minh Trail. This was a strategic axis that allowed troops and munitions to pass from the Vietminh during the Vietnam War, and it had been bombed incessantly during those years. It had not been easy to get authorization to visit this area and carry out paleontological research.

Alain Thiollier, the cultural counsel of the French Embassy in Laos, really made my task tremendously easier by transmitting my official request for a field mission to the Laotian authorities; he got us permission to go out in the field in the Savannakhet province, which was exceptional. I was able to set out for Vientiane on December 11. Two Laotian researchers, Phouvong Sayarath, from the Institute of Natural Resources and Energy of the Ministry of Science and Technology, and Houmphanh Rattanavong, who was in charge of social sciences for the Laotian Committee, were assigned to accompany me. Both had perfectly mastered the French language: Sayarath, a geologist, had graduated from the University of Rennes, and Rattanavong had been a cadet at the military school at Saint-Cyr. They were enormously useful in helping me to surmount the inevitable difficulties that we encountered during our trip. We reached Savannakhet by plane, and managed to find an all-terrain vehicle to rent in town. On Sunday, December 16, with the permission of the provincial governor, our trio, joined by one of Houmphanh's old Army companions, headed east on National Highway 9.

Now, it must be admitted that Highway 9 is little more than a succession of potholes. The passage of trucks loaded with logs and blocks of gypsum, as well as the absence of any sort of maintenance, had contributed to the complete deterioration of this major axis between Laos and Vietnam (as if the bombs weren't enough). Our all-terrain vehicle was no spring chicken, and our driver was in the habit of sticking his ear out of the window on a regular basis. When he heard a jerky sort of noise, which seemed to occur about every 15 minutes, he would stop the truck. It was nothing serious, he just had to screw the left front wheel back on. Okay, so three bolts were missing; there were still three left.

But there were other, more somber reminders of what decades of conflict had done to this country. As we passed by the old military base at Seno, I could see the rusty remains of cannons from the last Indochinese conflict, and when we stopped at the district's headquarters we were provided with a guard of five soldiers armed with Chinese AK-47 assault rifles. Our trio, which had become a quartet, was now augmented by a quintet of shock troops, whose weapons were nothing to sneeze at, and more than a little unsettling.

We passed by Muong Phalane, not far from where Hoffet had found the first remains of Laotian dinosaurs. About 20 km (12 mi.) east of this large village, we left Highway 9 and plunged into the forest on a pothole-riddled trail leading south. I was hoping to reach the village of Ban Tang Vay, about 30 km away (19 mi.), and it was only by overcoming some tremendous difficulties, as night was falling, that we managed to cover that distance. Time and time again, our vehicle would stall or get stuck in sand; roots and stumps barred our passage; a ford would have to be crossed. And so it was nearly one in the morning before we reached a little village that our inquiries informed us was right next to Tang Vay. We spent the night there, before reaching the objective of our trip early the next morning.

The sun came up over Tang Vay, a village of a few dozen very nice traditional teakwood houses, set on pillars and with shingle roofs. These dwellings were lined up between the forest and the rice fields, with their windows oriented traditionally from east to west, following the course of the sun in the sky. The villagers were attending to their usual chores and occupations: The children were bringing the water buffaloes to the river; the young girls were winnowing rice by beating the sheaves against a wooden plank to separate the grains from the stalks; the women were preparing the morning meal, husking the rice by using their legs to activate a lever attached to a large wooden hammer; the men were sitting under the larger trees, talking and smoking their pipes. Ducks, pigs, dogs, and roosters served as the official welcoming committee.

The elders and the chief of the village received us on the raised porch of the nicest of the village houses, and a long conversation ensued with my companions, the point of which was to understand why we had come. We were in a village of the Katang, a Mongo–Khmer minority whose language differed from Laotian, and my interpreters had their work cut out for them. But the villagers understood quite well that I was there for research . . . looking for the sacred buffalo. Some minutes went by; an old man shuffled up, delicate and reserved in appearance. He was introduced to me as Mr. Kommapa. I realized with a shock that he had known Josué-Heilmann Hoffet! At the age of 22 he had been assigned to accompany Hoffet in his ramblings through the Laotian forest, driving him from village to village and protecting him from ambushes and difficulties in the bush. Mr. Kommapa was now 76 years old, but he remembered perfectly the man that he still called "Mr. Minh." The name

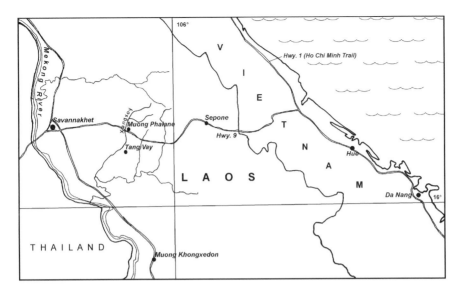

Figure 29. Map of the Savannakhet area, showing the Tang Vay fossil vertebrate localities and the footprint locality at Pha Lane. (After J.-H. Hoffet, modified)

undoubtedly had been given to Hoffet because he came from Vietnam, and perhaps because he worked for the Bureau of Mines.[2] Fortune was smiling on me, because Mr. Kommapa remembered very well where the bones had been discovered. He himself had carried a number of boxes containing the bones that had been packed for transport to Vietnam.

Trying not to let my excitement and impatience show too much, I asked if I could be taken to the place where these things had been found. The conversation between the villagers and my companions became more animated, that much was clear. It seemed that there were to be conditions. The bones belonged to the sacred water buffalo, and the evil spirits risked being disturbed if precautions were not taken. Forty-four years after Hoffet, these spirits still prowled around the village, and there was no question of violating village traditions. The last time that bones were collected there, it didn't sit well with the spirits, as the village elders still recalled: "They filled forty-three baskets (*kabung*) of bones from

[2] "Minh" is a common Vietnamese name; in addition, the pronunciation of the word "Minh" ("meen" to Anglophonic ears) is very similar to the French pronunciation of *mine*, which means "mine" in English.

an ox called *ngua sup'alaat,* the 'magnificent ox,' an ox that had a fore-head the size of a pillar (*ngua dai nôô sao*). When they began to dig, there was a storm, and flashes of lightning (*fa pha*), and it began to rain (*fôn lông maa*); this was a sign, a warning."

Clearly, there was no way around it; so many years after Hoffet first passed through, we were going to have to make a sacrifice. The local priest, a small man with quick eyes, was called, and I bought a pig for the price of ten thousand kips (about 100 francs, £10, or $17) – a sum I dutifully recorded in the budget for reimbursement to me, as the re-sponsible agent for the Museum. The sacrifice settled, villagers and vis-itors alike assembled for an excellent lunch of – yes, sacrificial pork, prepared according to traditional recipes: grilled, boiled, chopped and mixed with herbs and spices, marinated in fiery sauces. All of it was washed down with great glassfuls of *soum-soum,* the locally made rice brew. During the meal, one had to take care not to forget the ancestors who had gone on before; they were honored by spilling a few drops of alcohol on the ground through the floorboards. And all this had the sal-utary effect that the whole troop was authorized to take the forest road under the scorching sun of lower Laos, to follow Hoffet's trail.

As with many epic adventures, we started out as three and wound up reinforced by a band of about 20, armed to the teeth because a sec-ond group of militia, this time from the village, also accompanied us, bringing American M-16 assault rifles, grenades, rocket launchers, and other munitions. The official story was that this was to protect us in the forest. But I never expected to need this much firepower just to hunt dinosaurs! After an hour of walking, Mr. Kommapa led us to the small stream of Tiane Kang, to a place named Hue San Ok, which means "the place where the lance was planted." It was here that a sacred water buf-falo had once fallen in the stream while pulling a cart loaded with gold, which got bogged down; the animal, not being able to free itself, was dispatched with a stroke of the lance by one of the men who was with it. Today some people think that treasure is buried there, guarded by spirits. Anyway, fishermen say that in that place their nets are swept away, and that it is impossible to spend the night there.

In this sacred place, we reached a *pong,* or clearing, where Hoffet had collected his dinosaur bones. There was nothing more on the ground, not even the least bony scrap, not the faintest indication. We spent long hours in the forest prospecting every nook and cranny. It was hot, and our fatigue was catching up with us. A little discouraged, we were head-

Figure 30. "A small caudal vertebra of a Saurian" found near Tang Vay, Laos. (Photo by P. Taquet)

ing back to the road when suddenly, 400 meters from the village, right on the path, resting on a bed of leaves, was "a small caudal vertebra of a Saurian," as Hoffet himself described his first discovery. My adrenaline reserves skyrocketed in an instant: It was not just a dinosaur vertebra, but one from an iguanodontid, as its quadrangular form showed. If my hypothesis was right, there was a strong probability that the Laotian dinosaurs were not from the Late Cretaceous, as Hoffet thought, but from the Early Cretaceous, because this is when iguanodontids are found. The precise directions that Hoffet gave to his sites served us well. A few minutes later we found ourselves below the beaten path in a vast clearing. Near an unexploded mortar shell we found the lower end of a tibia, a few tail vertebrae, a nice neck vertebra, a few metatarsals, and numerous other fragments of a small iguanodontid! A few fragments of turtle shell and some crocodile bones rounded out our day's catch. We were delirious with excitement.

The return to the village was triumphant. All the children ran along with us and I took a souvenir photo of the male population of the village, who were delighted to pose with the "sacred buffalo" bones that we had brought back from our expedition into the forest.

The trip to Savannakhet and then to Vientiane proceeded without incident, after we had presented an account of ourselves to the provincial authorities in front of the portraits of Karl Marx, Ho Chi Minh, and the head of the Laotian state. In the capital, a press conference announced the new dinosaur discoveries of the Laotian–French team, to great public interest. A four-year agreement of cooperation in this research was drawn up, specifying that with the financial aid of the French Foreign Office, the specialists from the Museum of Natural History at Paris would come each year to study the Laotian fossils, train Laotian researchers and technicians, and help establish a new museum to feature these natural riches in Laos. I returned to France thinking about Josué-Heilmann Hoffet, about his joy in finding the first Laotian dinosaurs, about his efforts and his frustrations, and his isolation in the heart of these forests.

On January 17, 1991, at the request of Monsieur Fayard, the director of the Natural History Museum of the city of Grenoble, and the friends of the Museum, I traveled to Grenoble to give a lecture about the evolution of dinosaurs, which I called, "A voyage across the Mesozoic." This superb museum had just been renovated, and it concentrated mostly on the flora, fauna, geology, and paleontology of the Alps. At the end of the day I gave my lecture in front of a large audience, and I ended with some slides of our recent work in Laos, reminiscing about the work of the French geologist who had died for his country in 1945, after discovering the first Laotian dinosaurs.

After my talk, as the public shuffled slowly toward the exits, a woman, visibly moved, approached the front of the room. She thanked me for having paid my respects to Josué-Heilmann Hoffet, a man that the world had virtually forgotten . . . a man who had been her father. She was carrying a copy of a work that was none other than Hoffet's thesis on the geological formations of lower Laos. She had had no idea that I was going to talk about the dinosaurs of Laos. I had had no idea that I was going to meet Hoffet's daughter. I learned from her that Madame Hoffet and her four children, Anne-Line – the woman before me – Marie-Ève, Jean-Frédéric, and Catherine, would be happy to tell me more about the man whose discoveries had brought me all the way to Laos and back.

A few days later, I was back in my lab in Paris when I received an envelope from Marie-Ève with five very moving black-and-white photographs. There was a handsome portrait of Josué-Heilmann Hoffet, a fine man with a moustache and a straightforward expression, clear and vibrant. There were two field photos of our geologist in front of his tent near Sepone in 1935. Another was of his cot, his mosquito netting, and

EXTRAIT DES *COMPTES RENDUS DES SÉANCES*
DU CONSEIL DES RECHERCHES SCIENTIFIQUES DE L'INDOCHINE
Année 1942, 1ᵉʳ Semestre

Description de quelques ossements de Titanosauriens du Sénonien du Bas-Laos

par

J.-H. HOFFET

HANOI
IMPRIMERIE D'EXTRÊME-ORIENT
1943

Figure 31.   Cover page of the publication of the first dinosaur bones from Laos, by the French geologist Josué-Heilmann Hoffet.

Hoffet himself writing at his work table. There was also a picture of the forest clearing where the complete femur of the titanosaur was found, and a photograph of the femur, which is still in the geological collections of the University of Hanoi. Anne-Line expressed her joy in finding that "those who have left us cannot be entirely buried from our collective memory," and she shared her hope that she could go with us to Laos on our next campaign: She wanted to rediscover the memory of her father in the country that she had left at the age of nine.

Little by little, I set about gathering Hoffet's publications and documenting his interests, which ranged far beyond his travels in Asia as well as geology and paleontology. He was especially taken with the houses and customs of the mountain people of the Annamite range. He had sent back samples of fossil wood to the Museum of Natural History in Paris, and he had given the Museum of Man many examples of the typical household objects of Laotian villagers, as well as a very nice collection of black-and-white photos that he had taken during his 1933 travels in lower Laos.

The Museum of Natural History in Paris and the Museum of Man, thanks to the explorations and collections of many natural historians, geologists, geographers, and innumerable dedicated and generous amateur collectors, are the repositories of incomparable and extraordinarily varied things from all over the world. The history of this collection began in 1635; today it harbors a million fishes, 80,000 birds and mammals, 500 million insects, and 10 million plants. A great number of these are type specimens, used to standardize species when they were first discovered and named, and so they serve as an invaluable reference collection: Specialists must have recourse to them when they assess their own research collections.

Western researchers didn't explore Laos scientifically until relatively recent times. This mountainous country, landlocked amid Thailand, Cambodia, Vietnam, China, and Myanmar (formerly Burma), was first visited by a Dutch merchant, Gerrit van Wuysthoff, in 1641 and 1642. The first naturalist to visit it was the botanist and entomologist Henri Mouhot, in 1858; he was a native of Montbéliard, like Cuvier, the famous comparative anatomist and paleontologist. Mouhot died of fever in Laos, some miles from the ancient royal capital of Luang Prabang. In 1866, the frigate captain Doudard de Lagrée, along with his lieutenant Francis Garnier, started his military expedition; Dr. Thorel accompanied them as a naturalist and brought back 2,700 species of plants. But the fieldwork was primarily the last labor of an aged colonial infantry sergeant and employee of the military outposts, Auguste Pavie. From 1879 to 1895 he explored all parts of Laos along with his assistants, who included the geologist Counillon, and he published a great number of works. Counillon, who was first a teacher at the Saigon *lycée* and later the director of the nascent Geological Survey of Indochina, sent back to the Museum a collection of 431 plants. Near Luang Prabang he discovered a very interesting locality that contained some animals that used to be called "mammal-like reptiles," which today we would call synapsids. But the last of the great botanical collectors in Laos was Poilane, an agent of the Department of Waters and Forests who traveled all over the Indochinese Peninsula and sent more than 30,000 specimens back to Paris. This is why our Museum's herbaria are so rich in Laotian plants, which number nearly 50,000, and the greatest living specialist on the Laotian flora is a Frenchman, Jules Vidal, who has dedicated his life to studying the flora of this country.

The history of this whole part of Southeast Asia during the past 50 years has not been kind to humans, nor to other animal and plant life. The wars, with their bombings and defoliants, have added to the rapid deforestation caused by the slash-and-burn agriculture of the mountain people, as well as the intensive exploitation of the forests by the timber industry. These factors have reduced the forests of the Indochinese Peninsula to a mere shadow of their former glory, and have caused innumerable extinctions among the flora and fauna. This is why our precious Museum collections are so valuable: Through the years they have testified to the memory of the natural legacy of Laos, Vietnam, and Cambodia. They comprise a unique resource. And it was an unforgettable experience to be able to show the officials of the Laotian government this evidence of their legacy, which their scholars could now study as soon as diplomatic relations were resumed between France and this country with which it had been associated for so long (and in fact had never fought). The prospect of being able to construct, in collaboration with our own Museum researchers, a museum at Savannakhet that would house and display these natural riches, to exhibit in Laos the dinosaurs, plants, and other animals of that country, seemed to evolve as a matter of common accord among the people involved in the project.

According to the terms of our agreement, I prepared a second mission to Laos, this time providing for a full complement of scientists. We took along a number of our Museum's scientists, including Bernard Battail, a specialist in fossil reptiles, Jean Dejax, our paleobotanist, and Philippe Richir, our preparator in charge of casting, who had been with us in Morocco. We also brought our two Laotian friends, Phouvong and Houmphanh, along with two other colleagues: Phosykeo Tamvirith, from the Information and Culture Service of the Savannakhet province, and Bounxou Khentavong, from the *École normale no. 1* in Savannakhet. The goal of this mission, given our accomplishments in 1990, was to collect more fossil bones that were well preserved and still in place in their stratigraphic position.

A year later, we were standing once again in the village of Tang Vay. As Hoffet had done in 1936, we set ourselves up in one of the village houses for the duration of our stay. Our team had become larger with the addition of cooks, water carriers (whose buckets were made from the aluminum taken from American airplanes that had been wrecked during the war), and laborers, along with two armed guards who were assigned to accompany us. As soon as the welcoming ceremony under the beautiful village pagoda was over, we started work. For several weeks we shared the peaceful, pastoral life of the villagers, set to the rhythms of the chimes of the gong that were sounded by the young monks, the clink of the bamboo and teakwood bells that hung around the necks of the buffaloes as they meandered toward the streams, and the morning, evening, and nightly cries of dozens of roosters scratching beneath the piles of houses, including our own. In spite of all our efforts to consume as many of the foul creatures as we could at every meal, stewed or grilled, their numbers and noise seemed only to increase.

Against my hopes, the sites that we had visited in the previous year gave up only very few new elements: a few bones of iguanodontid dinosaurs, a carnivorous dinosaur tooth, and some fragments from turtles and crocodiles. But one fine morning, after we had already exhausted the fossil resources of the beds west of the village, a hunter came by to tell us, quite discreetly, that he knew a place in the forest where he had seen some things on the ground that were like those that we had collected. Remembering Hoffet's misadventure of walking for hours and finding nothing but some limestone blocks, I asked the man to go back to this place and bring us a few fragments of this "bone." He came back some hours later with a fragment of a very large rib that appeared to belong to a sauropod dinosaur. Our man was indeed a good observer, and surely a good hunter.

We decided to set off immediately to this new, promising site, located 2.5 km north-northeast of Tang Vay (1.5 mi.). A little winding path brought us into the heart of the forest and to a large clearing, an erosional basin where the sediments had been stripped bare by currents of water during the rainy season. In the bed of the little stream that crossed it, still embedded in its sandy matrix, rested a large rib indeed; around it the ground was littered with bony fragments. There was another "sacred buffalo" under our feet.

Beginning the next morning, we were out there working with all our equipment and material, thanks to our two all-terrain vehicles. And the work moved along very quickly: large tail vertebrae, metatarsals, tibia and fibula (lower leg bones) still connected to each other, the hind foot of a sauropod dinosaur of medium size – the tibia was about 90 cm (3 ft.) long – were disengaged, little by little. The preservation of the bones was excellent. As the days went by, all the parts of the hind foot were extracted from their matrix: We found the astragalus, a large rounded mass of bone, hollow at the near end for the tibia to fit into it; this is an interesting and characteristic piece that could help to tell us with which group among the sauropods we were dealing. The toes and three nice flat claws completed the bones of the hindlimb. As for the tail vertebrae, we saw quickly that they formed a connected string of decreasing size, and some days later we had excavated 35 of them. The tail was folded back on itself, long and whiplike. It is thought that these whiplike tails, as in *Diplodocus* (the animal that had so impressed President Fallières), might have been used to defend them from attacks by predators. Examining our Tang Vay sauropod's vertebrae showed me that in any event this was not a titanosaurid, one of the more derived Late Cretaceous forms, whose tail vertebrae were very characteristic: They had a cavity in the front and a rounded process in the back of the vertebral body, or centrum. The absence of this feature reinforced my suspicions that these deposits were from the Early, not the Late, Cretaceous.

The days rolled by quickly, interrupted by a narrow escape when Jean Dejax and Bernard Battail, on the road back to the village one day, had an unexpected encounter with a magnificent and aggressive cobra. Preferring extinct reptiles to live ones, they put their geological hammers to practical use and dispatched the poor cobra, who wound up in the evening's stewpot, to the general satisfaction of the company.

The last days of our fieldwork saw a visit by a French television crew, whose producer, Jacques Mény, was taken by the opportunity to convey the charm of the village of Tang Vay and the emotion connected with our rediscovery and exploration of the places and the people associated with the dinosaurs of Laos. A calf was sacrificed by the head priest in the presence of all the village men. Then, on Saturday, December 14, 1991, as we had planned, Anne-Line, Hoffet's eldest daughter, arrived at the village of Tang Vay to meet the villagers, especially Mr. Komma-pa. It was a simple meeting, intense and charged with the memories of her father, our predecessor.

More than 50 years after the first discoveries of the bones of the "sacred buffalo" of Tang Vay, we felt that we had renewed the strong links between the inhabitants of this tiny Laotian village and their visitors from a far-off land, France, who had long ago enjoyed such a privileged relationship with each other. We also felt that we were renewing the ties between the past and the present, following the course of history from the time of the great dinosaurs to the time of this little village, lost in the forests of Laos. But we also knew that our presence, our research, and our collecting would also untie some cultural knots: namely, those that had connected this place with the legend of the "sacred buffaloes." Would the children of the village of Tang Vay still fear the "sacred buffaloes" tomorrow? Or would they dream of becoming dinosaur hunters instead?

All our collections were brought to the capital of the province, and in the fall of 1992, during our third field mission, we set up an open-air studio for fossil preparation in the park next to the Museum of the Revolution at Savannakhet. There, before large, interested crowds, we extracted vertebrae, phalanges, and limb bones of the Tang Vay sauropod from their matrix. A little temporary exhibit was set up with signs in Laotian and French explaining the tail and hindlimb. The provincial authorities and thousands of schoolchildren came to see it. The whole thing was later put in a room on the ground floor of the Museum of the Revolution, in a beautiful old building that had been the official residence during the time of the French Protectorate. As word spread, more and more visitors came – some even from Thailand, who only had to cross the yellowish currents of the Mekong River, which bordered the park where the dinosaur was displayed.

We returned to Tang Vay for the third time, and the sacrifices of the previous year must have pleased the gods, because as soon as we arrived, one of the villagers, Mr. Phoun, told us that in the forest near his

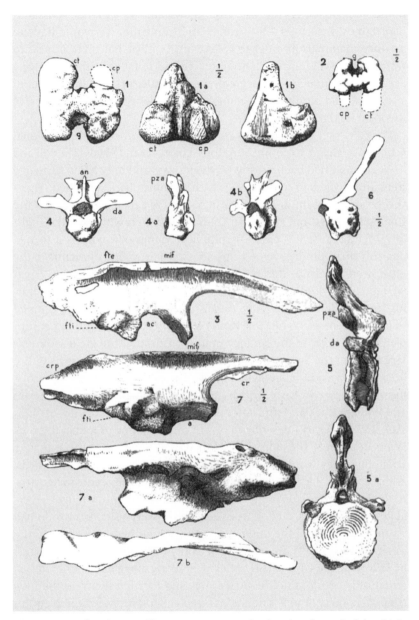

Figure 32. A plate from Hoffet's 1944 monograph, showing the end of the thigh, several vertebrae, and the ilia of *Mandschurosaurus laoensis*, an iguanodontid dinosaur from the Early Cretaceous of Laos.

house, not 500 meters (0.3 mi.) from the village, were more "sacred buf-falo" bones! It seemed a bit strange and ironic that we would have to sacrifice mammals in order to collect our reptiles, but there it was: To restore life to this ancient herbivore, we had to take the lives of sever-al cattle. Fortunately they were domestic cattle, because our Museum statutes would hardly have seemed compatible with the destruction of the local wildlife.

The new site, once again, was in a clearing formed by the erosional action of a stream, exposing burgundy-colored, sandy beds. The soil and its vegetation had been mostly washed away, and only a few shriveled trees and clusters of spiny bushes with a whitish bark remained. These shrubs are common in such clearings; they're called *gal tam pan* in the Katang language and *ton may* in Laotian, and their white wood is high-ly valued because it is easy to sculpt little things out of it. As it turned out, this small shrub gave us a clue about where to find dinosaurs in the great stretch of dry forests and clearings that made up lower Laos. When we plotted the locations of our fossil sites on the geologic map extend-ing from Phalane to Tang Vay, the distribution of points at first seemed to have no pattern at all. But Hoffet had observed long ago in his sup-plementary note on the geology of lower Laos that the great sedimen-tary basin of the Savannakhet province was breached across its entire breadth by a slight bulge, which geologists call an *anticline*. This term – introduced by two early British geologists, William Daniel Conybeare and William Buckland (the latter the discoverer of the first known dino-saur, *Megalosaurus*) – describes a sort of convex dome or fold in the rock layers in an area. The older geologic beds are found inside the anticline: Originally, they would simply have been beneath the flat, newer beds, as but geological forces pushed them up upward, they came to lie with-in the bulge. Here, the bulge extended from north to south over sever-al dozen kilometers, and on its eastern and western slopes, erosion had worn away some of the outer, younger geologic beds and exposed the inner, older ones. These exposed beds held the dinosaur bones. Thus, by following the edge of this anticline, looking along its flanks, and sys-tematically prospecting these clearings and locating them on the map (thanks to our portable satellite-assisted geographic positioning equip-ment), we could find other potential localities without wasting a lot of time and energy.

We started to excavate the new site, which was promising: Soon we had taken out five back and tail vertebrae, a pubis, an ischium, and a femur of another sauropod. The matrix was tough, and we had to use hammers and chisels on it; but it was very close to the village, and it occurred to us that if these bones could be preserved in situ, it would benefit the villagers of Tang Vay, who would in turn be responsible for the maintenance of the site. This would allow visitors not only to see the dinosaur bones displayed in an unusual setting but also to discover a traditional village. We could take the bones found at more inaccessible places in the forests to the museum in Savannakhet. We proposed this project to the village authorities, then to the provincial authorities, and finally to the French ambassador to Laos, all of whom had come to visit our digsite.

At the end of this third field season, we were talking in the field and on the way back about stopping at Phalane to have a look at the superb, flat, horizontal sandstone slabs that form the banks of the Se Sang Soi River, which you can see from the bridge that spans it at the entrance to the town. We thought that these slabs seemed likely to preserve dinosaur footprints, and we hadn't missed our guess: After a few minutes of looking around, Jean Dejax and I found a number of tridactyl prints arranged in three parallel trackways. They showed that three dinosaurs, probably ornithopods, had passed that way more than 100 million years before. It would be reasonable to think that the footprints had been left by the iguanodontidlike animals whose bones we had found at Tang Vay. Jean Dejax, examining the trackways in the oblique light of the setting sun, realized that two of the tracks were deeply impressed in the sediment, whereas the third hardly dented the substrate. The first two were found in a part of the slab that was covered with *ripple marks,* the long parallel stripes that form a light wavelike relief on the surface of sedimentary beds, caused by the flow of currents. The presence of these ripple marks told us that when these dinosaurs passed through, this stretch of water had had a slight current to it, and this is what had made the ripple marks. You can see the same thing today on sandy beaches, especially when the tide is going out, or on the edges of lakes and streams with mild, lapping waves. The third trackway being outside the ripple marks, Jean Dejax deduced that two of the dinosaurs had been walking along through the water while the third had been keeping his feet dry. Elementary, my dear Watson; Dejax would have been a useful member of Professor Challenger's crew exploring the Lost World!

These dinosaur trackways, the first discovered in Laos, could be found only for a few dozen meters along Highway 9, near the gates of Phalane. Their fortunate discovery completed the plan that we wanted to put into place for the future visitors to Savannakhet who might be interested in touring some of the natural riches of the area: We wanted to have a main museum in the city of Tang Vay, an on-site display with the bones at Savannakhet, and between them, an on-site display of the footprints at Phalane.

The fourth Laotian–French paleontological mission was dedicated to evaluating and conserving the field sites, continuing to excavate our sauropod bones and study the different parts of its skeleton, taking the measurements and features of the trackways, setting up the displays and their explanatory signs, and finally the official inaugurations by the Laotian Vice-Minister of Culture and the Provincial Deputy of Savannakhet.

In December 1993, Jean-Frédéric Hoffet, the son of Josué-Heilmann Hoffet, joined us at the Tang Vay site and helped with the dig under the watchful eye of Mr. Kommapa, the man who had guided and helped his father so many years ago. In the village, we had financed the construction of a traditional village house called the Maison Hoffet. Its main beams were laid just like those of the other village houses, in the presence of the head priest, the former cultural minister Alain Thiollier, and all the villagers. This house would accommodate all the future field crews and visitors to Tang Vay.

On Monday, December 20, 1993, the Laotian–French paleontological team was invited to the inaugural ceremony for a new French College at Vientiane, christened by the Minister of Foreign Affairs and the Minister of Education for Laos, as well as foreign ambassadors in the Laotian capital. This college is officially named the "Collège Hoffet," in honor of the pioneer of Laotian geology and paleontology.

In 1990, I had left to do research on Laotian dinosaurs, and I had found them along the paths of the Tang Vay forest; but I had also found the trail of Josué-Heilmann Hoffet, an extraordinary man who had impressed the inhabitants of that area and had distinguished himself by his discoveries of dinosaurs. After tying up the loose ends of this long story, whose first chapter began in the Cretaceous, I closed the book by traveling to Strasbourg to see the person who could tell me the most about Josué-Heilmann Hoffet – his wife. Along with Anne-Line and Jean-Frédéric, we leafed through her album of souvenirs, and the joy and emotion came flooding back, remembering the "sacred buffaloes" and the scorching sun of Laos.

ACROSS EUROPE WITH THE DINOSAURS

I
T SEEMS THAT everywhere we go in Europe today, our movie and TV screens have been invaded by a gang of American dinosaurs! And there's no doubt that America has been a tremendously rich hunting ground over the years. Besides, as I hope my own experiences in this book show, the more you look, the more you find – all over the world. But we shouldn't forget that the first representatives of these enormously popular reptiles were found in Europe. In every country of the European community there are a great number of sites, quarries, and outcrops that are important to the history of geology and paleontology. In fact, the paleontology of dinosaurs, as well as those of other fossil reptiles, was based upon discoveries made on European soil.

These discoveries and their histories are often overlooked these days, because newer events have supplanted them in the popular consciousness. For example, today the village of Maastricht, in Holland, is rightly celebrated in the eyes of the public, mainly because the Treaty of European Union and Cooperation was signed within its walls. It is less well known that long ago, the city also gave its name to the last geological stage of the Mesozoic Era, the Maastrichtian – the stage that saw the final disappearance of the dinosaurs. Oxford is naturally well known for the elegance of its university campus, and for the charm of its little river, so dear to the author of *Alice in Wonderland*. It is less well known that the striking neo-Gothic building that houses its Museum of Natural History displays the remains of the *Megalosaurus*, the first dinosaur that was ever described. From Maastricht to Oxford, from Solnhofen to Aix-en-Provence, you can map out a kind of paleontological ecotourism, a scientific itinerary that lets you follow step by step the often neglected

■

backroads that were main highways for the fossil hunters of the eighteenth, nineteenth, and twentieth centuries. These roads offer you the opportunity to learn how some of the basic concepts of earth science were formulated, and to see how this knowledge emerged, step by step. In pointing out some milestones for you here, I'd also like to show you that some of the fits and starts in the history of the peoples and countries of Europe are intricately linked to the episodes and fortunes of scientific discoveries.

In the early 1820s, some bones were found in Great Britain in some ancient deposits near Oxford; they were rumored to belong to a new and bizarre kind of reptile, never before seen on Earth. But before this discovery could be properly understood, scientists had to prove that some species had once lived in remote times but were now extinct. This was the achievement of a young and brilliant French anatomist, Georges Cuvier, who demonstrated elegantly and definitively that certain fossil bones belonged to great reptiles that had long ago populated the land and seas, but had since disappeared.

Cuvier was 26 when he was called to Paris, to the National Museum of Natural History, to take over the duties of the anatomist Mertrud. After successfully completing his studies at the Caroline University of the duchy of Württemberg at Stuttgart, he had spent several years in Normandy as a tutor to the Protestant family of the Marquis of Héricy. Immediately noticed for his gifts as an anatomist, he was able to give free rein to his talents – and to his ambitions to be like his senior colleagues such as Lamarck and Geoffroy Saint-Hilaire. He arrived in 1795, joining a prestigious establishment that had just survived the torments of the French Revolution, and that had passed in 1793 from the status of the Royal Garden of Medicinal Plants to that of the National Museum.

Now, the year before Cuvier arrived at the Museum, the collection of the Cabinet of Natural History received an exceptional specimen that was also a spoil of war: the fossilized skull of the Great Animal of Maastricht. This specimen, whose meaning baffled the entire scholarly world of the time, was to play a great role in the history of paleontology.

St. Peter's Mountain at Maastricht, Holland, is a hill that dominates the city, and its tablelike relief has resulted from the erosion of the Maas (or Meuse) River. The horizontal geologic beds of St. Peter's Mountain were deposited a little more than 65 million years ago, at the end of the Cretaceous, and they are the result of the sedimentation of billions of tiny organisms with calcareous shells. This calcareous rock, more solid

Figure 33. A portrait of Georges Cuvier by the English painter Henry William Pickersgill, done a year before the death of the illustrious naturalist. Cuvier wears the cross of the Legion of Honor and the cross of the Commander of Württemberg.

than the chalk of the Paris Basin, has been used throughout the centuries for construction material, and since the Roman Era the hill has been riddled with subterranean quarries. The historian Pliny alluded to their existence in 50 A.D. In the eighteenth century, mining was very active, and many hundreds of kilometers of tunnels pierced the mountain in all directions.

In 1770, the quarrymen discovered some enormous jaws, implanted with conical and lightly curved teeth, from the skull of a gigantic animal in one of the chalky banks about 500 ft. from the main entrance. The discovery caused a great sensation, and Dr. Hoffman, a retired German military surgeon and amateur of natural history, examined the fossil remains. By judiciously spreading some money around, he got the workers to carry the limestone block containing the skull to his residence. The Dutch anatomists from the Teyler Museum in Haarlem, Pieter Camper and his son Adrien, each gave their opinions on the identity of this animal armed with such great jaws. The father supposed, based on the size of the bones and on the presence of associated sea urchins, that it was a kind of marine animal, a whale; the son supposed that they were dealing with a gigantic lizard, but no known lizard was capable of attaining such dimensions. Others imagined that the bones belonged to a crocodile.

But Hoffman was not able to enjoy this spectacular specimen for long. Canon Godin, who owned the land below the quarry, saw fit to exercise his rights; he sued Hoffman and won, thus coming into possession of the Great Animal of Maastricht. Godin placed the fossil skull in a glass case in the middle of his garden so that the public could admire these antediluvian remains. Unfortunately he in turn could not enjoy his acquisition for long. The French Army occupied the city of Maastricht in 1794, because it was well situated at a strategic crossroads, and annexed the province to France. Godin hid his treasure from the armies of the French Republic, but the scientific commissioners charged with drawing up the inventory of objects to seize as spoils of war demanded the return of the celebrated fossil skull. To recover it, General Pichegru promised 600 bottles of wine to whichever soldiers succeeded in getting their hands on the scientific treasure. Within hours, the piece was unearthed and transported directly to Paris – where it's still found today.

When the skull of the great animal arrived in Paris, Cuvier had just finished some very detailed anatomical comparisons between the living elephants of Asia and Africa and the fossil mammoths and mastodons. This work had allowed him to confirm that certain species were extinct – that they had disappeared from the face of the planet. Cuvier could be confident in writing that "so many remains of unknown animals have belonged to creatures from a world before ours, to creatures destroyed by some revolution of the Earth." (He really did think that catastrophes were the principal agent of biological change through time on Earth, as we shall see in the next chapter.) As for the reptiles, Cuvier wrote – in a letter of 1788 to his German friend Pfaff, whom he had met at the university at Stuttgart – that they had begun their existence far earlier: "Their debris fills more ancient formations, and the naturalist is obliged to pursue their remains into even deeper beds."

Figure 34. The skull of the Great Animal of Maastricht, the mosasaur, collected from the underground quarries of St. Peter's Mountain, as displayed today in the paleontological gallery of the Museum in Paris. (Photo by D. Serrette)

Cuvier thus took the study of the Great Animal of Maastricht into his immense program of describing an ever-growing number of known fossil bones, and he used the same methods of comparative anatomy that had served him so well in his study of the elephants.

There remains for me to discuss the most celebrated, the most gigantic of all, the one that has brought on the most controversy, having been previously taken for a crocodile, for a saurian of some other genus, and even for a whale or a fish. . . . Monsieur Faujas, after his return, published a work entitled *Natural History of St. Peter's Mountain*, in which he engraved beautiful figures of these objects. He accepted the ideas in vogue at Maastricht and constantly refers to our animal as a crocodile. . . . But Monsieur Adrien Camper convinced himself that they came from a particular kind of saurian reptile that has some affinities with the monitors (the varanids) and others with the iguanas. . . . We will show today that Monsieur Adrien Camper is the only one who has really understood the characteristics of this animal.

In a dazzling demonstration, Cuvier went on to show the force of his reason and to resolve this paleontological enigma. He showed that the structure of the teeth was highly instructive: The animal from Maastricht had solid teeth that were completely fastened to the jaw; this eliminated it from the crocodiles, whose teeth are implanted in cylindrical sockets without being fused to their walls, and the whales, whose unfused teeth are implanted in fluted grooves. Moreover, the lower jaw of whales is composed of only one bone, whereas the Maastricht animal's jaw is closer to that of monitors, which comprise several bones. So it came down to a choice between monitors and iguanas. Monitors have conical teeth, whereas iguanas have lance-shaped teeth with jagged edges; the animal from Maastricht has teeth like those of monitors and not like those of iguanas. Finally, the bones of its palate are very different from those of the palate of a fish, but very similar to those of the monitor. An examination of its backbones showed that they were dealing with an aquatic animal. Cuvier could thus conclude that the Great Animal of Maastricht was a marine animal close to the monitors:

Doubtless it will seem strange to some naturalists to see an animal so surpass in dimensions those to which it is closest in the natural order, and to find its remains in marine beds, inasmuch as no saurian appears to live in salt water today; but these singularities are hardly considerable in view of so many others that offer us numerous monuments of the natural history of the ancient world. We have already seen a tapir the size of an elephant; the *Megalonyx* provides us with a sloth as big as a rhinoceros; why should we be surprised to find, in the animal from Maastricht, a monitor as big as a crocodile?

And to drive home his demonstration and impress his reader, Cuvier, who did not lack a flair for the dramatic and who knew how to showcase his discoveries, ended with these words:

> But what is important to notice is this admirable constancy of zoological laws, which do not fail in any class or in any family. I had not examined either backbones or limbs before I occupied myself with the teeth and jaws, and one sole tooth was enough to reveal all this.

The Great Animal of Maastricht was named *Mosasaurus,* the saurian from the Meuse, by the Englishman Conybeare in 1822, and it was another Englishman, Gideon Mantell, who erected the species *Mosasaurus hoffmanni* in 1829. The description of this great extinct lizard comforted Cuvier in the idea that Earth had been subjected to waves of successive extinctions, of revolutions of the globe. Today, the mosasaurs that populated the Upper Cretaceous seas, contemporaries of the dinosaurs, are still considered close to the monitors, and Cuvier's diagnosis has never been supplanted.

Our visit to the subterranean quarries of St. Peter's Mountain at Maastricht merits a detour. This will allow us to admire the beautiful succession of limestone beds of the Maastrichtian. On the wall of one gallery, a grand sculpture representing a mosasaur commemorates the discovery of 1770, while the Dutch guide, turning to his French visitors, uses the occasion to remind us that the celebrated skull was taken in 1794 by the French Army and transported to Paris. The history of this abduction is well known to the inhabitants of the city, and I have some personal experience of this myself. One Saturday in the spring of 1993, while I was working in my office, which adjoins the paleontological gallery of the Museum, I suddenly heard a song with patriotic accents, rising up from the second floor of the great exhibit hall. Intrigued by this unusual display (for our Museum, at least), I glanced into the gallery to find, to my surprise, about 30 students bedecked with T-shirts displaying the coat of arms of the city of Maastricht, standing before the skull of the mosasaur . . . singing the Dutch national anthem! So it goes with fossils, as with paintings, sculptures, or any other precious objects: Their histories are sometimes tightly linked to the stormy relations between nations. Nevertheless, in adopting the Maastrichtian as one of the reference stages in the worldwide geologic time scale, geologists and paleontologists showed as far back as 1849 that they were Europeans who were thoroughly convinced that the rocky beds that they study have no borders.

Even as it was becoming clear that giant marine reptiles had once lived but had completely disappeared from Earth, there was increasing evidence in the nineteenth century that the terrestrial realm had similarly been inhabited by huge reptiles. But it was not until 1842 that some of these terrestrial reptiles were segregated into their own group, the dinosaurs. Before 1842, there had indeed been discoveries of fossil terrestrial reptiles, but their interpretations were somewhat difficult.

As it turned out, Normandy was very rich in fossil vertebrates, and early on it provided amateurs and casual collectors with a lot of Mesozoic vertebrate remains. In January 1776, the first written trace of these discoveries appeared; it was published in the *Journal de physique*. It all goes back to the Abbé Dicquemare, an astronomer and zoologist who was born in Le Havre in 1733. He was combing the beaches of Normandy pursuing his favorite pastime, collecting and studying marine organisms, when he came across some fossil vertebrae at low tide. He collected these at the base of the cliffs in the Auge region, between the Touque and the Dive rivers, across from the cliffs known as the "black cows." Dicquemare put these observations in the *Journal de physique*:

In the beds of a bank of lead-colored rock, full of a considerable quantity of more or less petrified fossil shells, whose species, except the scaly oysters, are never found in neighboring waters, some years ago I perceived some rather large masses that, when examined at close range, seemed to have a texture and substance like those of bone. Most of them were shapeless, but turning my attention to those that had this characteristic, it was not difficult to notice that each piece had an envelope of tissue that differed from that in the center: I thus recognized the appearance of bone.

These petrified bones were called *osteoliths* ("bone-stones") at the time. Rather than constructing strong hypotheses about the origin of these bones and whence they came, the Abbé Dicquemare preferred simply to describe them, adding: "It is in communicating that one may hope to add to the sum of our understanding about the interior of the Earth and the sudden or successive revolutions that it must have undergone." The Abbé described these fossils and recognized that one of them was a left femur. Unfortunately he didn't depict this piece, but his description was extremely precise: An ammonite was adhering to the bone in the area of the lesser trochanter (a bump near the head of the femur for muscle attachment), and the dimensions of the femur were twice those of a human femur. This could not have been the femur of a fossil crocodile; it is quite probable that the abbé had collected the femur of a dinosaur.

In 1800 Cuvier, who had traveled in Normandy some years before, described a new species of fossil crocodile in the *Bulletin de la société philomatique de Paris*. In 1808, he described these remains in more detail in the *Annales du Muséum*. They had been part of a collection made long ago by the Abbé Bachelet, a naturalist from Rouen. Cuvier remembered Dicquemare's ancient discoveries, and he described these bones from the areas around Honfleur and Le Havre, comparing them to those of the gavial, the long-snouted fish eater of the Ganges River in India. For Cuvier, all these vertebrae could be divided into two distinct lots: those that were similar to vertebrae of living crocodiles, and those that belonged to completely unknown crocodiles. Today we know that these latter vertebrae belonged to dinosaurs. Cuvier made drawings of these vertebrae, which he assigned to a crocodile, in his own hand for a publication in 1808; they became the first dinosaur bones to be illustrated in France. His description and drawings were reproduced in 1812 in the first edition of a major work that assured his reputation: *Recherches sur les ossemens fossiles de quadrupèdes ou l'on rétablit les caractères de plusieurs espèces d'animaux que les révolutions du globe paraissent avoir détruites* (*Researches on the bones of fossil quadrupeds, wherein are established the characteristics of the many species of animals apparently destroyed by the revolutions of the globe*).

When his work became well known, Cuvier traveled to Great Britain with his wife and daughter. He was accompanied by his artist, Charles Laurillard, a close friend who, like Cuvier, came from Monbéliard. Laurillard's impressions of the trip were as follows:

As you know, we took a trip to England, which was very interesting for Monsieur Cuvier and his family. We saw some beautiful things in the way of natural history, especially fossils. Their collections, however, cannot compare to ours. The Royal College of Surgeons has some wonderful preparations made by Hunter, and some skeletons apart from these, but the English anatomists, to say nothing of the naturalists, are not strong in zoology and comparative anatomy, so they don't even realize the value of their riches. In general, the scientific establishments in England are virtually nothing; the government is only interested in the art of making money, which is carried to perfection in this country. They make money out of everything: collections of paintings, natural history cabinets, views of monuments; indeed, everything costs money in England, everything comes back to money, and everything is related to money.

Laurillard's judgment was a little severe; scientific life was particularly active across the Channel at that time. Cuvier went to Oxford in 1818 to meet "one of the men who honors geology by his precise and consistent observations, and by his most constant resistance to unsupported hypotheses," the Reverend William Dean Buckland. Buckland showed him the vertebrae of a large fossil animal that some quarrymen had just collected from the little town of Stonesfield. "Professor Buckland had made this great discovery some years before, and I saw its pieces at his house in Oxford in 1818; I even drew some of them." Cuvier realized that these vertebrae closely resembled those that he had from Honfleur: They had unusual cavities on their sides; likewise, the transverse apophyses, or side processes, of the vertebral arches arose from four prominent ribs that formed a pyramid-shaped base. But Cuvier didn't take his observations further, and it's not clear why. He knew at the start of his trip about the existence of large terrestrial fossil reptiles. He had at hand some strange vertebrae (in fact, those of a carnivorous dinosaur, though he didn't realize it) of Jurassic age, like those that Buckland had. But these were the discoveries of his British colleagues William Buckland and Gideon Mantell, and they didn't really understand the full importance of their discoveries either. It fell to Richard Owen to discover the final clues to the mystery, and to understand the uniqueness of these reptiles – which he named dinosaurs.

William Dean Buckland was born in 1784 on the border between Dorset and Devon, in a region particularly rich in fossils of all sorts. He entered Corpus Christi College at Oxford, where he earned his diploma in 1804; in 1819 he was named Professor of Geology there. He was fascinated by the remains of the past, which he collected and studied with great care. Buckland knew that all these fossils were the remains of organisms that had once been alive, and he tried to reconstruct their appearances, their diets, and their ways of life. In this respect, as Phillip Powell, the current curator of Oxford's museum says, he could really be called the pioneer of paleoecology. Buckland was curious about everything, not just geology and paleontology; he was also involved in many political and social questions. When he became canon of Westminster, he had them install toilets. He tried to better the lot of the British by introducing many new foods, like cornmeal.

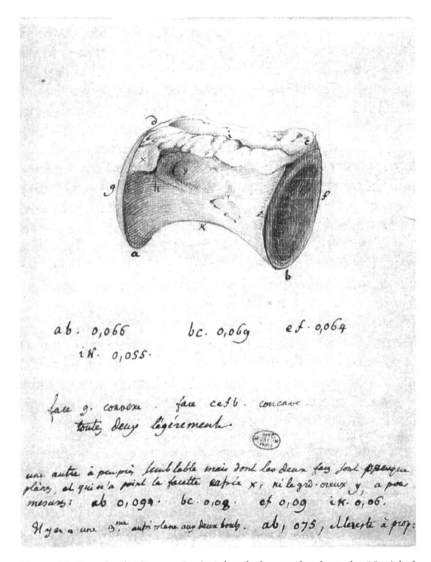

Figure 35. Drawing in Georges Cuvier's hand of a vertebra from the "Gavial of Honfleur" – really a dorsal vertebra from a carnivorous dinosaur – that the anatomist published in 1808. (Photo courtesy of the Central Library of the Museum)

The study of the fossil bones from Stonesfield fascinated him. After Cuvier's visit, he described in detail all the pieces that had been collected – vertebrae, hip bones, a lower jaw with teeth, ribs, elements of the fore- and hindlimbs – and he knew he was dealing with a reptile. Buckland was intrigued by the probable posture of this animal, which looked like a lizard but whose forelimbs were much shorter than the hindlimbs. And a lizard whose dimensions must have reached a height of two meters – a lizard the size of an elephant! In 1824 Buckland named this great lizard *Megalosaurus,* or "giant lizard." He sent Cuvier the plate from his publication, which had been drawn by Mary Morland, his future spouse. Buckland authorized Cuvier to use the figures of the bones of *Megalosaurus* that he had sent him, along with the figures of Cuvier's Honfleur vertebrae, to put together a new plate for the second edition of Cuvier's work on fossil bones. In return, Cuvier sent Buckland casts of the distal ends of a tibia and a fibula, an astragalus, and a calcaneum (the bones surrounding the ankle), as well as the end of the pubis of another large fossil reptile discovered at Honfleur, which seemed to belong to the same type of giant lizard. These casts are still found today in the collections at Oxford, under Phil Powell's care.

The discoveries of fossil reptiles in Great Britain multiplied. Another personage entered the scene: Gideon Algernon Mantell, a country doctor at Lewes, in Sussex. Mantell, a collector of rocks and fossils from the surrounding countryside, was preparing a substantial work on the fossils of the South Downs, among which were several bones found in the ancient beds of that region. He described with particular care the discovery of a certain number of teeth belonging to an unknown animal.

The circumstances of this discovery deserve a brief mention. It is said in some accounts, though not all, that during a visit to one of his patients, Mary Ann Mantell, his wife, was walking along a road that was flanked by piles of rocks that were to be used to shore up the embankment. From one of the piles, she picked up a rock that had some shiny objects encased in it. These curious objects were fossilized teeth, of a genus that was then unknown, and Mary Ann showed them to her husband when she returned. Mantell was able to find out where the rock came from; it was from a quarry near the village of Cuckfield, and Mantell succeeded in getting other specimens of these mysterious teeth. In 1820, Cuvier had just returned from Great Britain; he'd heard of Mantell's discoveries and the imminent publication of his work. Cuvier contacted Mantell through a common friend, Mr. Greenough of London, whom he asked to get him a copy of this book on the fossils of the chalk hills of Sussex. On July 28, 1821, our naturalist country doctor began a correspondence with the French anatomist and paleontologist, sending him some plates from his upcoming work, as well as some specimens. Throughout the summer of 1822 Mantell discovered other teeth, and not knowing what animal to assign them to, he sent some to Cuvier, probably by way of Charles Lyell, the great founder of geology, who was going to Paris. On June 20, 1824, Cuvier wrote to Mantell, giving his opinion on these fossils, which were then unknown to science.

These teeth are certainly unfamiliar to me; they are not at all from a carnivorous animal, but even so I believe that they belong to a reptile, by virtue of their simple form, the serrations on their edges, and the thin coat of enamel that covers them. From their outside appearance, they could be taken for the teeth of fishes like the Tetrodons, or the Diodons; but their internal structure is much different than these. Are we not dealing with a new animal here, a herbivorous reptile? And even more so since today, among terrestrial mammals, it is among the herbivores that one finds the largest animals; so it could have been among the ancient reptiles. Inasmuch as they were the only terrestrial animals, why couldn't the largest of them have eaten plants? A part of the large bones that you have belong to this animal, the only one of its kind so far. Time will confirm or deny this idea, because it's inconceivable that we will fail to find a part of the skeleton connected to jaws bearing these teeth one day. It is always the final thing that one must research with the most perseverance.

One can easily see, reading this letter, that Cuvier gave a very astute answer, implying the possibility of fossil reptiles of great size. Then, on July 9, 1824, Mantell sent some other teeth to Cuvier, so that he could include them in the second edition of his *Ossements fossiles,* as he had done with Buckland's specimens. Cuvier supplied some commentary on the teeth Mantell had discovered:

It is not impossible that they belong to a saurian, but a saurian even more extraordinary than any that we know. What gives them a unique character is their transverse wear on both the tip and the barrel, as in herbivorous mammals; so much so that the first tooth that was presented to me being found in this state of wear, I had no doubt that it came from a mammal; it even seemed to resemble the molar of a rhinoceros, which, given where it came from, would have upset all my ideas about the relationships of the fossil bones with their sediments. . . . Only when Monsieur Mantell sent me a complete series that were more or less worn was I completely convinced of my error.

History records – on the British side, at least – that Cuvier was completely wrong in attributing these teeth to a rhinoceros. Reading his correspondence in 1983, I was able to do justice to the other side of the story, and to show that, even though Cuvier had not been close to the concept of the dinosaur, nevertheless he did not ignore the importance of this new herbivorous reptile, which he knew represented a completely new kind of animal.

But Gideon Mantell continued to pursue his quest to discover the owner of these teeth. His colleague Samuel Stutchbury, working as a naturalist for the British Museum, brought to his attention the troubling resemblance between the teeth of iguanas, the herbivorous South American lizards that Stutchbury had recently seen in the wild, and the teeth of Mantell's animal from the Cuckfield quarries. This is why Mantell took up his pen once more to share this important news with Cuvier:

Monsieur,

Since I last had the honor of addressing you, respecting the herbivorous reptile, whose teeth and bones are found in the sandstone of Tilgate Forest, in this County, several fine specimens of teeth, and numerous fragments of bones of an enormous magnitude have been discovered. Some of these teeth are figured in the annexed plate, and as they are more illustrative of the dentition of this animal, than those previously submitted to your notice, I was anxious to transmit this plate to you.

You will I have no doubt immediately perceive how closely some of the teeth resemble those of the Iguana of the Barbadoes; in fact so striking is this analogy that I have ventured to propose the name of *Iguanosaurus* for the fossil animal, as indicating the resemblance it bears to the recent one. The recent skeleton with which I compared the fossils, is in the Museum of the Royal College of Surgeons . . . the size of the fossil teeth is however from four to six times that of the recent ones . . . the metatarsal probably belongs also to this animal.

If it will not be trespassing too far on your indulgence, and encroaching too much on your invaluable time, may I Sir respectfully solicit the honor of being favored with your opinion on this subject. Be pleased Sir, to honor me by accepting of my most distinguished consideration.

With the most profound respect, I am he

Gideon Mantell

Mantell finally decided to give the name *Iguanodon* to the new terrestrial reptile discovered in the Tilgate Forest sandstones; the name means "iguana tooth." A year after the description of *Megalosaurus*, the *Iguanodon* entered the dinosaurian history books. Mantell pursued his researches and his labors, and some years later he had made some solid progress in understanding these great fossil reptiles. In 1834, quarrymen from the city of Maidstone succeeded in extracting a good part of a *skeleton* of *Iguanodon* from a large block of rust-colored sandstone. For the sum of £25, which was pretty steep in those days, Mantell acquired the block. This important discovery allowed him to describe this relatively complete skeleton and to reconstruct the animal, the first time this had been done for a dinosaur. Mantell represented the *Iguanodon* with a horn on its nose; we know today that this pointed bone is actually one of the *Iguanodon's* thumbs.

But it was with Owen that the dinosaurs finally made their name, as well as their remarkable entry into the world of extinct species. Richard Owen, who eventually became the first Director of the British Museum (Natural History), understood not only that these giant reptiles with their columnar limbs were different from other reptiles and more similar to elephants on this point; he also understood that their barrel-shaped rib cages seemed more mammalian than reptilian. He supposed that dinosaurs had four-chambered hearts and warm blood like mammals do – which was adventurous for the time, like many revolutionary hypotheses that were bound to be forgotten before they could be properly debated by dinosaur specialists. Owen put forward his share of daring ideas, but he was normally a firm conservative and a convinced "fixist": He vehemently opposed Darwin's ideas on the evolution of species, which he thought were subversive and dangerous.

Owen moved in the highest Victorian circles. He used his influence with Prince Albert, Queen Victoria's husband, to propose the financing of the three-dimensional reconstruction of the first known dinosaurs, *Megalosaurus, Iguanodon,* and *Hylaeosaurus,* for the first international exposition in modern European history: the Crystal Palace exhibition, which was to move from its original site in London's Hyde Park to the more expansive Sydenham Park, south of London – where they can still be found today. These reconstructions were conceived by the artist Waterhouse Hawkins and installed in the reassembled and enlarged Crystal Palace, a great building of glass and metal. Bricks, cement, and maple branches were used, and on New Year's Eve of 1853 a most unusual inauguration took place. A banquet was held *inside* the reconstruction of the *Iguanodon,* whose broad back had not yet been completed. Twenty revelers took their seats inside this impromptu restaurant, and under the portraits of Cuvier, Buckland, Mantell, and Owen they proposed toasts to the glory of the dinosaurs – and Queen Victoria, of course.

The dinosaurs of the Victorian Era are still standing under the foliage of Sydenham Park, and even if these nineteenth-century restorations seem wildly inaccurate today, their simple charm and the way in which they're presented are really worth a visit. So are the small villages of Stonesfield and Cuckfield, and Mantell's former house at the foot of Lewes Castle.

Naturally, since these historic times, dinosaur discoveries have become more and more numerous, whether in Britain, France, or other countries, and whole volumes have been written to take inventory of all the dinosaurs discovered since Buckland's description of *Megalosaurus*. Among these discoveries, two might capture our attention: One is quite old and took place in Normandy, the other is relatively recent and occurred in Surrey in 1983.

One morning in July 1835, Jacques Amand Eudes-Deslongchamps, professor of natural history on the faculty of science at Caen and secretary of the Linnean Society of Normandy, was alerted by one of his colleagues that some bones of very large dimensions could be found in a block at a construction site on the Bayeux road in one of the suburbs of Caen. It was urgent to get there right away, because children were playing on the blocks and removing the bones, which excited their curiosity. Eudes-Deslongchamps hurried to the site on the Bayeux road, but could only survey the damage: The bones were destroyed. Nevertheless, he carefully collected the remaining fragments. He tried to buy the block, but it was Sunday and the man in charge was gone. This delay was fatal, because the next day more bones were missing. Going door to door, Eudes-Deslongchamps managed to recoup some pieces, including the upper end of a femur, a large number of phalanges, and what he thought were portions of the metatarsals. This block had come from a quarry in Maladrerie, a village located a quarter-league away (a bit less than a kilometer, or five-eighths of a mile). Eudes-Delongchamps got there and acquired some poor fragments from the quarry workers, but he also learned about another block with bone that had been taken out and brought to the city. Four or five days later the faculty chemistry technician put him on the track of this second block. This time he had to run down to Basse-Rue, near the port. "The ossiferous block, sized, carved, and trimmed, was just about to become the lintel of a window.

Figure 36. The *Iguanodon* statues in Sydenham Park, south of London, constructed by Waterhouse Hawkins in 1853. A banquet was held inside one of the unfinished statues on New Year's Eve of that year.

I managed to acquire it." This time, Eudes-Deslongchamps recovered seven fairly complete vertebrae.

But the professor's adventures were not over. A week later a worker from the quarry came to his house carrying a large number of loose bone fragments in his handkerchief. The man had taken out these fragments with a sledge hammer. The results were not pretty, but Eudes-Delongchamps was still able to construct a dozen vertebrae out of them, with the help of some glue and lots of patience.

Three weeks later, Eudes-Deslongchamps found yet another quarry worker at his door. This one was all in a sweat, riding a horse that was the most ill-tempered old nag that he had ever seen in his life. This man brought him a new package of bones from the same quarry. "But it was a pity to see the state of this debris, how it had been mistreated; most of it was powder." These bones came from another block 60 feet long. Eudes-Deslongchamps hurried over to the site and succeeded in taking away 12 or 15 small blocks from it. "So I was finally in possession of my treasure; but Good God, what a state it was in!"

By persistence alone he managed to reassemble the bones of a single animal: "Some twenty tail vertebrae, a humerus, a radius, an ulna, two phalanges of the hand; a femur, some fragments of the tibia and fibula, some ankle bones, metatarsal fragments, and a great number of toe bones; and many ribs, some of which had extraordinary shapes that led me to infer that they were placed amid the abdominal muscles." Eudes-Deslongchamps, despite all these difficulties, had gotten hold of the bones of a carnivorous dinosaur. And as he observed, it had ventral ribs that we now call *gastralia,* which formed a sort of basket under the animal's stomach, perhaps to protect the abdomen. This is a diagnostic feature of carnivorous dinosaurs.

Eudes-Deslongchamps set to work describing this extraordinary animal from the middle Bathonian beds (Middle Jurassic), about the same age as those of the Stonesfield megalosaur. The bones belonged to a large animal some 25–30 ft. long, or about 10 meters. Eudes-Deslongchamps knew about the recent discoveries by Cuvier and Buckland, but he thought that his animal was considerably different from *Megalosaurus,* and so he named it *Poekilopleuron bucklandii.* The genus name, composed of two Greek words (*poikilo-* meaning different or variable, and *pleuron* meaning rib, lung, or chest) alluded to the variable form of the ribs.

Jacques Amand Eudes-Deslongchamps left to posterity a magnificent monograph published in Caen in 1837. His work on *Poekilopleuron* is a model: He described each bone in detail and lingered over the particular structure of the ventral ribs and their positions, trying to find similar structures in living reptiles. His work had hardly been published when William Dean Buckland announced himself at Caen. This was quite an event; he was accompanied by Lady Buckland and their two children. Their trip had started off badly because their steamboat ran aground in the Orne on the way from Le Havre, and the illustrious voyagers had to travel two leagues (about 6 mi.) in the August rain to reach Caen. As a Normandy naturalist wrote much later in a humorous vein, Buckland "was under the influence of the tradition of the universal flood, which had for so long rained down maddingly on the study of the Quaternary and on the question of the ancient age of Man." (The allusion, of course, is to the story of Noah's Flood. The Bible had been interpreted by some clerics to limit the age of the Earth to about six thousand years, though naturalists were beginning to think that it was much older.) One of Buckland's friends wrote that, until Buckland's work, ev-

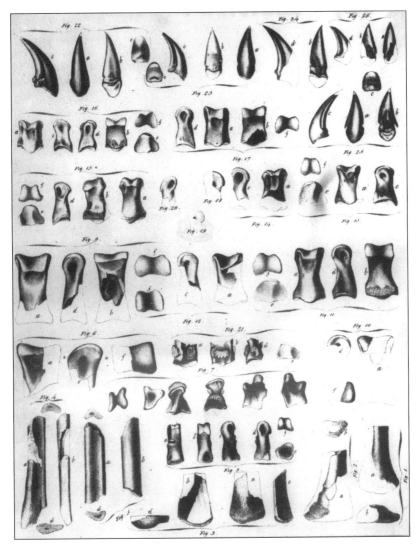

Figure 37. The bones of *Poekilopleuron bucklandii,* a carnivorous dinosaur from the quarries near Caen, as restored and described by Jacques Amand Eudes-Deslong-champs in 1837, and destroyed by the bombardments of the Second World War.

erything about the story of the Great Deluge had been obscure, but that Buckland had made it "as clear as mud!"

At Bayeux, Buckland made the obligatory visit to the famous tapestry; he attended the public meeting of the Antiquary Society of Normandy, then called on Eudes-Deslongchamps in order to admire his paleontological collections and to visit the localities around Caen with him for a couple of days.

I was lucky enough to find a few letters from Eudes-Deslongchamps some years ago. One of them, unpublished, is of interest to us because it tells us about the very different reactions of Buckland and Eudes-Deslongchamps to the object of this historic visit. Eudes-Deslongchamps was very close to Étienne Geoffroy Saint-Hilaire, professor at the Museum in Paris and an excellent anatomist devoted to the skull anatomy of fossil and recent crocodiles. On August 21, Eudes-Deslongchamps wrote a letter to Geoffroy Saint-Hilaire from which the following paragraphs are taken:

Monsieur and most honored master,

I have waited to thank you for all your kindness to me, for the note of well-wishes by which you have honored my memoir, which, following your desire, has been communicated to the Academy.

As for my work, you judge it much too favorably; it is a work of conscience and long effort, and that's all. . . . Those more clever than I could make much more of these materials than a modest mason like myself, using only the strength of my shoulders, working at the edges of the palace of paleontological science of which you are the primary architect.

The several trivialities that are found in the first part of my memoir could have easily been avoided, and the entire first part could even disappear without harm to the rest. In writing it, I was under the influence of the memory of the enormous material difficulties that I had to overcome in order to put my bones in a condition to be identified and preserved. . . .

I had the opportunity to see Mr. Buckland at Caen; I was very eager to show him the bones of my great saurian and to ask his view on its identity and on its possible identity with *Megalosaurus*. He did not think that the vertebrae belonged to it; he thought that the femur, the tibia and the bones of the foot might belong to *Megalosaurus*, being similar to some fragments that he collected at Stonesfield and referred to this saurian; the ribs seemed marvelous to him, he had seen nothing like them except in plesiosaurs. He thought that my humerus and the two bones of the forearm were not at all like those of *Megalosaurus*; in short, he concluded that my bones belonged to two different species of reptile. But on this point, he was clearly in error. The circumstances in which I collected them could leave no doubt that they belonged to the same individual, and my examinations, which I have repeated a thousand times on all the pieces, confirm it indubitably. The disproportion between the forelimb and the hindlimb is the main reason that Mr. Buckland gave, and the possibility that two reptiles could be found buried together. But he did not know, or forgot, that all our crocodilians from Caen have this disproportion, even more pronounced. It is quite possible that my *Poekilopleuron* is none other than *Megalosaurus;* I have said so often enough in my memoir. I will be the first to applaud their identity when it is demonstrated; but the idea that my bones belonged to two species of animals that happened to be preserved together just to make a poor naturalist look foolish is too far-fetched for me to admit.

This letter shows how much Eudes-Deslongchamps stood by his ideas; we can also see that it was very difficult to have a clear, well-defined opinion about the appearance and identity of these animals in the early days of dinosaur discoveries. And even though this dinosaur was to be renamed *Megalosaurus poekilopleuron* by the German paleontologist Friedrich von Huene, its discovery and description by the Normandy naturalist introduced us to one of the most interesting dinosaurs ever found on French soil.

Unfortunately, however, it seems that evil spirits continued to toy with the fate of *Poekilopleuron,* which had been so painstakingly reconstructed by Eudes-Deslongchamps. In 1944 an Allied air raid bombed the city of Caen, his university, and his paleontological collections, and once and for all reduced his bones to bits (we must say "his bones," as he did in his letter).

Has *Poekilopleuron bucklandii* receded once and for all into the dungeons of history? It's not certain. We have an excellent description and figures of fine quality; but we also have casts of it. Yes, as it turns out, Eudes-Deslongchamps made at least one plaster cast of each of the principal pieces of the skeleton from Maladrerie, and they're preserved in our Museum in Paris! The holotype of the animal has disappeared, but we have what is called the "plastotype." And so we're ready for the next time someone finds specimens of *Poekilopleuron* in the rocks of Caen – or anywhere else.

Spectacular discoveries like this one by Eudes-Deslongchamps are not very common. However, occasionally completely new dinosaurs with surprising anatomical features are dug up from beds that have hidden them for many millions of years. Take, for example, my own experience in Niger in 1973, at the Gadoufaoua locality, whose Cretaceous fauna I had been studying since 1965. I had the occasion to collect the front ends of two jaws, and they both had the same characteristics: Each was robust and slightly spoon-shaped in dorsal view; seven tooth sockets were preserved on the left dentary row, as on the right one. The two branches of the jaw were tightly fused along their midline to form a solid beak. These pieces certainly belonged to the lower jaw. They could not be premaxillas, the bones at the front end of the upper jaw, because they bore no trace at their back end showing sutures for contact between the premaxillas and maxillas. What's more, the dentary rows were continuous at least up to the seventh tooth, and it was well known that the premaxilla in crocodiles and dinosaurs had only three to five teeth – and we were certainly dealing with the front end of the jaws. The reptilian nature of these pieces was further assured by the presence on each piece of thecodont teeth, that is, teeth that are implanted in the sockets but not fused to them. So these were mandible or lower jaw fragments, but from very strange mandibles, because the left and right dentary rows were separated by a bony longitudinal wall that I had never seen on any reptile mandible. The only reptiles that had a mandible whose anterior branches could be closely fused to form a long and narrow snout belonged to crocodiles; but in that case, the two edges of the dentary bones, in the lower jaws, are separated by a horizontal plate, which conforms to the shape of the palate in the upper jaws. A vertical bony wall between the dentary rows isn't present in crocodiles. Each of the front ends of these jaws bears several teeth. On one specimen the teeth are smooth, straight, and flattened, with edges that have small serrations, like on a steak knife, to help slice through flesh. The other specimen has striated teeth ornamented with long grooves. What sorts of animals were these, anyway, with such strange mandibles in a long, narrow snout?

I asked myself that question for several years. Once in a while I would take these two jaw fragments out on my desk and stare at them, trying to solve the puzzle. Carnivorous dinosaur jaws didn't run alongside each other so narrowly, and their *symphysis,* where the two jaws connect in front, was U-shaped or V-shaped, whereas this symphysis was Y-shaped, with a long fused piece sticking out in front. But then, looking through the dinosaur literature, I found that a mandible much like those from Niger had been described in 1915 by a German paleontologist named Stromer, who called his animal *Spinosaurus aegyptiacus.* His mandible piece had been collected along with some vertebrae that had very long spines; they came from a level within the lower Cenomanian (the first geological stage of the Late Cretaceous), from the *jebel* called El Dist in the oasis of Baharija, west of the Nile River in Egypt. So there was at least one other dinosaur that had this type of specialized jaw!

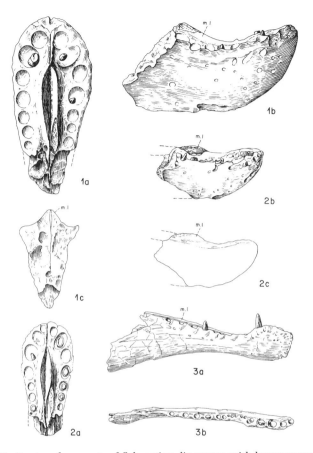

Figure 38. Dentary fragments of fish-eating dinosaurs, with long narrow snouts, collected from the Early Cretaceous of Africa. 1 and 2 are from the Gadoufaoua locality in Niger; 3 is *Spinosaurus aegyptiacus,* from the Baharija locality in Egypt, described by Stromer in 1943 and unfortunately destroyed during the Second World War. (After Taquet 1984)

Comparing his fragment with the dentary of a tyrannosaur, Stromer had pointed out that the front part of the mandible in *Spinosaurus* was probably very narrow: "It seemed to me in any case that the two branches of the mandible were not very divergent, by reason of the short symphysis, and that they belonged to an animal with a long, narrow snout." But Stromer's material didn't have the left and right dentaries in articulation with each other, so he may well have underestimated the length of the symphysis.

I considered this along with what I could gather from the jaw fragments in front of me – the Early Cretaceous bones from southern Morocco and southern Algeria – and came up with a conclusion that was
not especially orthodox. I could confirm that on the African platform
during the Early Cretaceous there was an extremely specialized lineage
of carnivorous dinosaurs with long, narrow snouts. This elongate shape
started me thinking about what they must have eaten. The closest points
of comparison were the many groups of narrow-snouted crocodiles and
other reptiles who had lived in the past. This long-snoutedness, or *longi-
rostry,* usually accompanies a fish-eating habit, as we can see today in
the gavial, the long-snouted, sacred crocodile of the Ganges River that
eats only fishes – and apparently has no fear of the crowds of people
that bathe in its river. So there may well have been fish-eating dinosaurs!

After some years of hesitation, I decided to take the plunge and offer my ideas at the annual meeting of British paleontologists, who were
congregating that year at the University of London. I remember that my
talk was heard by all the eminent specialists there in polite silence. No
one asked me any questions afterward. And though none of my colleagues rose in indignation to challenge this heretical hypothesis, none
showed any particular enthusiasm for my new evidence either. So, in
1984, I published a note in the *Comptes rendus de l'Académie des sciences*
entitled, "A Curious Specialization of the Skull of Certain Carnivorous
Dinosaurs of the Cretaceous: The Long and Narrow Snout of the Spinosaurids." My conclusions were as follows:

Nothing prevents us from supposing the existence of piscivorous theropod dinosaurs. All the Cretaceous localities from the African platform abound in fish
remains and it is tempting to imagine spinosaurid dinosaurs fishing along the
rivers or the edges of lakes like herons and storks. Spinosaurids offer a fine
example of the variety of adaptations in dinosaurs, and this discovery can be
added to other recent ones of new theropods, such as *Deinonychus, Deinocheirus,* and *Syntarsus.* This diversity of theropod form only serves to emphasize
the need to abandon the classic but overly simplistic division of theropods into
only two groups: the Carnosauria and the Coelurosauria.

It would be tremendously interesting to be able to examine the jaw
fragment that Stromer discovered at the oasis of Baharija. At the time,
the specimens he collected were taken to Munich. But, as I found out,
all the Egyptian material met the same tragic fate as the *Poekilopleuron*

bucklandii of Eudes-Deslongchamps: It was destroyed in the bombing of the Second World War.

Things remained more or less like that until a discovery in Great Britain made people talk seriously about fish-eating dinosaurs. In January 1983, an amateur paleontologist, William Walker, discovered an enormous pointed claw in a clay quarry from the Early Cretaceous in Surrey. This claw belonged to a carnivorous dinosaur of more than respectable dimensions, to put it mildly, and soon a crew of paleontologists from London's Natural History Museum was on the scene. They spent three weeks collecting the bones of a skeleton taken from a matrix of enormously hard claystone, impregnated with rust. They had tremendous difficulty removing the bones from the blocks at the laboratory because the rock was so resistant. But little by little, the well-preserved bones of a magnificent carnivorous dinosaur emerged.

My colleagues and friends in the Department of Palaeontology at the Natural History Museum, Alan Charig and Angela Milner, began to study the bones of this animal, which was one of the best discoveries of dinosaurs in Britain in over a century. They soon saw that this big carnivore had some astonishing features, including dentaries and maxillaries that showed that the animal had – who would have predicted it? – a long, narrow snout. This splendid specimen, exhibited today in London, was named *Baryonyx walkeri;* the genus name alludes to the large, pointed (*bary-*) claw (*-onyx*), and the species name alludes to its discoverer. It was placed in a new group of theropod dinosaurs, the Baryonychidae, which differ from the Spinosauridae by virtue of the absence of the long spines on the dorsal vertebrae. Of course, in considering the life-style of *Baryonyx,* my colleagues reasoned exactly as I had in supposing the existence of piscivorous dinosaurs. And in the article in *Nature* in which they announced their find, Alan Charig and Angela Milner, with the utmost elegance and collegiality, reproduced my statement on this point – in French, even! – that I had published in my note to the Académie: "it is tempting to imagine spinosaurid dinosaurs fishing along the rivers or the edges of lakes like herons and storks." But they added: "We independently conceived a similar idea for *Baryonyx,* though as a quadrupedal predator crouching on the bank rather than a biped stalking through the shallows." The great claw on each hind foot might have been used to spear fishes as bears, especially the grizzly, do in streams today.

The unexpected discovery of completely new dinosaurs in the Early Cretaceous of Africa and Europe clearly showed (if further evidence were necessary) the diversity of dinosaurian adaptations throughout the Mesozoic, and testified to their continually dynamic evolutionary patterns . . . so dynamic, in fact, that we now know that during the Late Jurassic, a little more than 140 million years ago, certain small carnivorous dinosaurs gave rise to birds!

It was in 1860 that a bird feather was found on a plate of lithographic limestone in a quarry in Bavaria. The German paleontologist Hermann von Meyer immediately understood the importance of this discovery: The ancient age of the beds in which it had been found by a quarry worker showed that birds appeared at a very distant point in prehistory, and that they had been contemporaries of the dinosaurs. In 1861, von Meyer proposed to call the animal represented by this feather *Archaeopteryx lithographica* ("the ancient feather from the lithographic limestones"). Then, in the same year, in the Langenaltheim quarry near the village of Solnhofen, a worker discovered the owner of this feather: a superb skeleton of an animal that looked reptilian but had beautiful feathers all along its wings and tail. The quarry workers of the time were poor, and the country doctor of the village, Karl Häberlein, was often compensated for his medical services in fossils. Häberlein sold his splendid piece to the British Museum in London for the sum of £700, which he used to provide his daughter with a substantial dowry. The *Archaeopteryx* arrived in London in 1862. Richard Owen was director of the museum at the time, and he quickly saw the importance of this fossil, which had features of reptiles and birds in the same skeleton. Owen was a fixist when it came to ideas about species: He did not think that species changed, or transmuted, into others. But such a fossil could only attract tremendous interest and curiosity, especially given the fact that Charles Darwin had just published his major work, *On the Origin of Species,* in 1859. The new fossil was coming at just the right time, it seemed, to furnish science with a particularly spectacular example of evolution. It was a perfect missing link, showing the transition from reptiles to birds – though Owen did not see it quite this way.

The *Archaeopteryx,* with its jaws full of small teeth, its long tail of fine vertebrae, so different from the "parson's nose" of the tails on today's birds, its forelimbs with three long claw-bearing fingers – these were all reptilian features. But on the other hand, the presence of a wishbone (which corresponds to fused clavicles, or collarbones) and, above all, the presence of feathers unquestionably made it a bird. But who was the reptilian ancestor of *Archaeopteryx*? Who were the ancestors of birds?

These questions have been debated broadly since 1862, and crocodiles, pseudosuchians (basal archosaurs from the Triassic), ornithischian dinosaurs, and saurischian dinosaurs have all been proposed as ancestors in their turn. But for many years no consensus could be reached, despite the great number of publications on the subject.

In 1876, a second specimen was discovered in the same region of Bavaria, not far from the little village of Eichstätt. This one was better preserved than the first, and this time Karl Häberlein's son Ernst put it up for sale at an exorbitant price: 2,600 German Marks. The German authorities had firmly decided that this exceptional specimen would not fly away to a foreign land. Finally, thanks to a substantial grant from the industrialist Werner von Siemens, the fossil stayed in Germany and was taken to the Humboldt Museum in Berlin. Since then, five other specimens of *Archaeopteryx* have been found in the lithographic limestones of Bavaria, the most recent in 1992. One of them has mysteriously disappeared from the private collection in which it was kept, and an international alert has been issued among researchers in an attempt to recover it.

But before any *Archaeopteryx* specimen was discovered, before this missing link became an object of lucrative commerce, before it was the object of a tug-of-war between various proponents and adversaries of evolution, a small carnivorous dinosaur was exhumed from these lithographic limestones. It was about 1850, in the region of Riedenburg Kelheim, in Bavaria. This little animal, about 70 cm long (28 in.), died on its right side and was perfectly preserved, with its neck stretched back over its body. This gracile little hollow-boned dinosaur was named *Compsognathus longipes* in 1861 by the German paleontologist Andreas Wagner.

During the 1970s, the American paleontologist John Ostrom, whom we met in an earlier chapter, was studying the carnivorous dinosaur *Deinonychus*, which he had collected in the Early Cretaceous of Montana. Looking at some strange features in its skeleton, he was drawn to reopen the question of anatomical comparisons between dinosaurs and birds. He went to Germany and examined the only specimen of *Compsognathus* in Munich, as well as the various specimens of *Archaeopteryx* that were then known. His conclusions were definite: More than 20 skeletal characters were shared by *Archaeopteryx* and small carnivorous dinosaurs such as *Compsognathus*. The number of characters was too high to be the result of evolutionary convergence; only direct relationship could explain why so many were shared. Suppose that the hypothesis that birds descended from little carnivorous dinosaurs is the most probable one. The contemporaneous *Compsognathus* and *Archaeopteryx* would be very close cousins with a close common dinosaurian ancestor. Today Ostrom's hypotheses are largely accepted in the scientific community, though they have sparked many debates. But when a second discovery of *Compsognathus* was announced in France in 1972, more than a century after the first specimen was found, John Ostrom packed his bags and headed for France. I set him up with my colleagues at the Natural History Museum of Nice, where the specimen was then kept.

This French *Compsognathus* came from Late Jurassic limestone beds on the plain of Canjuers, overlooking the gorges of the Verdon. These beds were of the same age as those in Bavaria, and the environment was also a lagoon bordered by coral reefs. The specimen had been preserved much as the one from Bavaria had been: The animal was resting on its right side, the neck was thrown back over the body, and the tail was broken around the seventh tail vertebra, probably from postmortem contraction of the muscles. But this specimen was larger: It reached nearly 120 cm (4 ft.) in length. This splendid, well-preserved skeleton is the most beautiful dinosaur skeleton ever found in France.

Quarry workers had mined the Canjuers locality for nice slabs of limestone to use as ornamental stone; but at the same time it also yielded the skeleton of a large crocodile 3.6 meters long (nearly 12 ft.), two complete turtles, dozens of fishes, small sphenodontids (the *Sphenodon* lives today only on islands off New Zealand and is related to lizards), and a flying reptile (pterosaur). However, the plain of Canjuers was also the site of a large military installation that was in the habit of firing barrages of artillery. You might imagine that this could jeopardize fossil collecting in the area, which was now exceptionally interesting to us. Léonard Ginsburg, one of our paleontologists at the National Museum of Natural History in Paris, knew the area well, thanks to his extensive work in geologic mapping. One of his geologist colleagues, Georges Mennessier, had learned of the discovery of fossils in the area and gone out to the site. The Museum, the paleontology laboratory, and our director at the time, Jean-Pierre Lehman, negotiated with the Ministry of Defense. With the influence of General de Gaulle's former aide-de-camp, they agreed to permit study and excavation of the site during breaks in the shelling – which was nice for us, because usually they were firing 144-mm shells. This agreement is still in effect today, and with the help and cooperation of the army, paleontological research takes place each summer at Canjuers.

When Canjuers became a military encampment, the business that had been excavating the quarry was taken over and the excavations ceased. The superb collections of fossils from this locality, including *Compsognathus*, were displayed in the hangars of the former owners. When I was named director of the paleontology laboratory at the Museum, one of my first tasks was to ensure that the magnificent fossils did not meet the same fate as the first *Archaeopteryx*, sold to the highest bidder in a foreign museum. In the course of a particularly tough negotiation, thanks to the help of a French banking group, Crédit du Nord, I succeeded in acquiring the entire collection, including the second specimen of *Compsognathus*, an animal rarer than *Archaeopteryx*. A few years ago this collection was placed on exhibit in the paleontological gallery of the Museum, where it can be seen today.

The *Compsognathus*, a distant cousin of *Archaeopteryx*, gives us an excellent idea of the kinds of animals that were the probable ancestors of birds. *Archaeopteryx* resembled it so closely that when the fifth specimen of the first bird was discovered in 1951, it was taken for a juvenile *Compsognathus*, because the traces of feathers on the specimen could be seen only in very low-angled light. This specimen, the size of a crow, is found today in the Jura Museum at Eichstätt, a magnificent castle that displays the fauna and flora that lived with the first birds, and it is the pride and joy of its curator, the delightful Dr. Günther Viohl. This is another beautiful stop on your itinerary discovering the dinosaurs of Europe.

The long and the short of it is that if you remove the feathers from an *Archaeopteryx*, you get an animal very close anatomically to a little carnivorous dinosaur. Feathers, which are made of keratin, are only the transformed scales of reptiles. This is so true that if you inject retinoic acid under the skin of the feet on chicken embryos, they'll grow feathers there where normally there are scales. The idea that the pigeons on our town squares and the chickens in our henhouses are the descendants of dinosaurs has left more than one person scratching his head. But since the discovery of *Archaeopteryx*, other links have reinforced the chain that led little carnivorous dinosaurs to fly. One of the most recent is Spanish, and was found in the center of Spain at Las Hoyas, in the province of Cuenca. It was studied by my colleague José Luis Sanz. This bird, more recent than *Archaeopteryx* (it dates from the Early Cretaceous), was called *Iberomesornis*. It is more advanced on the avian line than its

Jurassic predecessor: It's a mixture of characters from primitive birds and modern birds, and its discovery is of great importance in understanding the transformation from dinosaurs into modern birds.

So Europe has played a primary role in the discovery of the first great fossil reptiles, then in the discovery of the first dinosaurs, and finally in the discovery of the first birds. Many new discoveries in European countries add to our understanding of the origin of species and enrich museums that were already well-endowed with great specimens. To show the crucial role of Europe in this history, I got together with the film director Jacques Mény to assemble a show for Arte, the cultural production company, that was dedicated to the places, discoveries, and people that figured in the history of dinosaur paleontology since 1824. This tour took us to Maastricht, Bernissart, Brussels, Stuttgart, Tübingen, Eichstätt, Berlin, London, Oxford, Stonesfield, and Cuckfield, then on to the province of La Rioja in Spain, and to Portugal, south of Lisbon, to film and interview my Spanish and Portuguese colleagues on the dinosaurian treasures of their countries.

Our expedition ended up in France, in Provence, at the foot of Mount St.-Victoire, in the middle of the countryside that was so dear to Cézanne. This is where the first dinosaur eggs were found, and the locality is so famous that my American friends call it "Eggs in Provence."[1]

At the foot of Mount St.-Victoire, the slightly elevated beds of the Aix basin expose continental red beds in which the fossilized eggs are very abundant, and there are at least six different kinds of eggshell. These beds, which are of Late Cretaceous age, succeed each other through several layers of the last two stages of the Mesozoic Era, the Campanian and the Maastrichtian (at least 80 to 65 million years ago). They were discovered in 1869, but it was two scholars from Aix-en-Provence, Raymond Dughi and François Sirugue, who were in charge of the Natural History Museum of the city, who brought these exceptional sites to light through their collections and their labors. The largest eggs had a diameter of 20 cm (8 in.) and were attributed to the sauropod *Hypselosaurus priscus*. The bones of this titanosaurid are relatively abundant in the region, particularly at a site located at the foot of the little village of Fox Amphoux, nestled in the Var. Here, crews from the Museum carried out many successful field campaigns, thanks to the help and understanding

---

[1] This is a pun on "Aix-en-Provence," the famous fossil area. In French the word "Aix" is pronounced much like the English word "aches," not a far cry from "eggs."

of the landowners, Monsieur and Madame Icard. But there was no evidence of the identity of the dinosaurs who had laid the eggs, and the custom is to give different names to eggs (for example, *Megaloolithus mammilare* or *Megaloolithus siruguei*) than one would to the bones of the egg layers, just as we do for footprints. More than 65 million years ago, the area was a broad floodplain where great rivers emptied into vast lakes. The dinosaurs chose this area to come and bury their clutches of eggs in the sandy banks of these rivers. It's thought that after sudden storms, a large part of these nests were flooded; the eggs couldn't hatch, and this is why we can find them by the dozen today.

The quality of their preservation is such that the eggs are choice museum specimens. Unfortunately the sites are threatened. Forest fires, such as the one that devastated the flanks of Mount St.-Victoire in 1986, have made the soil very fragile, and erosion is rapidly destroying a number of the richest egg sites. If this were not enough, vandals and commercial fossil dealers contribute to the serious degradation of these exceptional sites, destroying or removing the eggs. Fortunately, under the pressure of local collectives, particularly the office of the general counsel and his conservation agency, the value of the land is now being fully appreciated, and a decree has recently been published in the *Journal officiel* identifying the most important sites. A project of protection and conservation should permit us to preserve this great paleontological resource for future generations; at the same time it will become an open-air museum, where the public can see these eggs and nests preserved in place.

Taking advantage of our visit to the egg sites at St.-Victoire, I brought our little crew of filmmakers to the eastern edge of the Aix-en-Provence basin, where the small community of Trêts lies, so that they could visit the owners of a superb estate set on the flank of Mount Olympus. Monsieur and Madame Michelon were waiting to show us the fruit of their harvest – not of grapes, but of some less conventional collections that they had made on their land. Following the terrible fires that had ravaged the area and had nearly destroyed their beautiful Provençale mansion, erosion had exposed some sandy clays on the neighboring hillside. Out one day to collect mushrooms, Madame Michelon had instead found some nice dinosaur-tail vertebrae and numerous bone fragments scattered on the slope. After carefully gathering these unusual objects and noting their position, she had the good sense to notify paleontologists of her discovery. Her brother had been kind enough to come to

Paris to show me some samples and photos that clearly established their importance.

Our visit to the site confirmed the importance of the find, and after a survey by two of my colleagues in the summer of 1993, we returned with a small crew of paleontologists from the Museum in the spring of 1994 to start quarrying the fossil-bearing bed. It took a few days before we fully understood that Madame Michelon had found what I think is the best site for Late Cretaceous dinosaurs that has ever been found in France. In these sandy and clayey beds, dozens of well-preserved dinosaur bones from four species (a theropod, a sauropod, an ornithopod, and an ankylosaur) are piled up together, as well as two species of crocodile and two species of turtle. All these species are represented by skull bones, vertebrae, limb bones, and girdles. All the bones are complete, with their joint surfaces in excellent condition, which is not generally the case in Late Cretaceous beds. We also had the good fortune to receive visits from two notable guests: my friend Jack Horner, the American paleontologist well known for his discoveries of dinosaur eggs in Montana, his *Tyrannosaurus* specimen, and his technical assistance on the films *Jurassic Park* and *The Lost World;* and my friend Dan Grigorescu, the energetic dinosaur specialist from Transylvania, Dracula's haunting grounds in Romania.

And so it seemed that I had come full circle. After the sands of the Ténéré, the shrubby undergrowth of the *sertão*, the slopes of the High Atlas, the steppes of the Gobi, the forest of Tang Vay . . . I had rediscovered the scrubland of Provence, the scents of thyme and lavender, the colors of the wild iris and the rose, the song of the nightingale, and far off, the splendor of Mount St.-Victoire. It only goes to show how much there is left to discover, sometimes right in our own backyards. Other dinosaurs, and other people, and other places, will help paleontologists add new pages to the great book of the history of dinosaurs in Europe.

CHAPTER NINE

■

THE RISE AND FALL
OF THE DINOSAURIAN
EMPIRE

A S THE GREAT eighteenth-century English historian, Edward Gibbon, judiciously reflected, rather than asking why the Roman Empire disappeared, we would do better to wonder why it lasted so long. The same could be said for the Age of Dinosaurs: 150 years after the word *dinosaur* first entered our language, the articles dealing with their extinction have never been more numerous, and they seem to increase weekly.

The disappearance of the last nonavian dinosaurs, a little less than 65 million years ago, is really one of the enigmas of paleontology. It has stimulated great floods of ink and dozens of hypotheses that have tried to propose a satisfying explanation for this event. There is no question that these strange reptiles fascinate the scientific world as much as the general public, including writers, artists, and cinematographers. In the early days of paleontology, dinosaurs were reconstructed in both scientific and popular realms in what we might see today as an extremely simplistic way. Dinosaurs were summarily described as necessarily stupid and malevolent beasts, poorly adapted to their environment and therefore doomed to extinction. Today they are considered survivors, a group of reptiles whose longevity through geologic time, diverse range of adaptations, and vast geographic distribution were second to none among the vertebrates, including mammals.

The dinosaurs first appeared at about the border between the Middle and Late Triassic, around 220 million years ago, as a lineage that split from among other archosaurian reptiles. Archosaurs are reptiles that share several unusual evolutionary features, including a skull opening between the socket for the eye and the naris, which houses the nostril.

■

Among the diverse lineages of archosaurs, a small branch called the Ornithodira had a new way of hooking up the ankle: The nearer row of ankle bones (the astragalus and calcaneum) were tightly linked with each other and with the tibia and fibula, respectively. The farther row of ankle bones and the foot bones to which they articulated moved against this unit in a hingelike way. This was quite different from the configuration taken up by crocodiles and many other groups, in which the astragalus was hooked up to the leg bones and the calcaneum was hooked up to the rest of the foot, so that the ankle joint essentially crosses the ankle. In the Ornithodira, which includes the dinosaurs and pterosaurs, and their relatives, the hingelike ankle goes along with a hindlimb that is oriented parallel to the body axis, so the animals tend to put one foot in front of another, close to the body midline. In the crocodiles and their relatives, the ankle splays the foot outward as the animal walks, and there is more of a sideways component to the gait.

Lagosuchus was a little animal from the late Middle Triassic of Argentina, about rabbit-sized (its name means "rabbit crocodile"), with a long tail and a total length of about 50 cm (20 in.). It is one of the first known ornithodirans, and probably very similar to what the earliest members of this group, including the ancestors of dinosaurs, were like. Its forelimbs were much shorter than its hindlimbs, and it probably got around on two legs most of the time. At the end of the Triassic, about 200 million years ago, the dinosaurs had evolved from within this ornithodiran lineage and diversified rapidly into two major groups, the Saurischia and the Ornithischia, that spread over all the continents (which were then connected as the single supercontinent, Pangaea), including Antarctica.

Do the dinosaurs form a monophyletic group? That is, are they descended from the same common ancestor – something like *Lagosuchus,* perhaps? – or did they spring from different groups of ornithodiran ancestors? After all, their pelves and various other features are somewhat different: Saurischians retain the three-pronged pelvis of other reptiles, whereas in Ornithischians the original pubis has migrated backward to lie alongside the ischium. In the 1970s some workers, including Robert Bakker, Peter Galton, and John Ostrom, began to speak strongly in favor of dinosaurian monophyly, which had been doubted for decades. Then, in the 1980s, younger workers began to apply the newer methods of cladistic analysis, which had been developed by the German entomologist Willi Hennig in the 1950s, to the question of dinosaurian monophyly. In 1984 and 1986 the American Jacques Gauthier

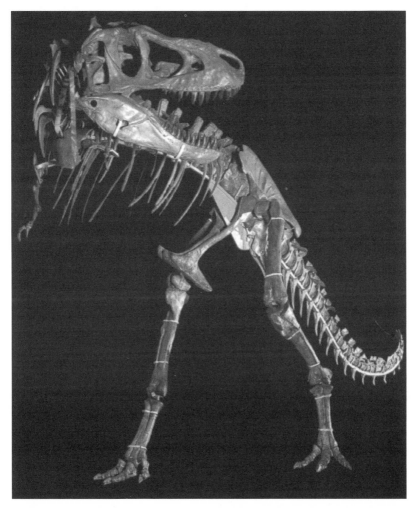

Figure 39. The complete skeleton of *Tarbosaurus bataar,* as displayed in Paris in 1992, from the collections of the museum of Ulan Bator, Outer Mongolia. (Photo by D. Serrette)

supported Bakker and Galton's conclusion that the dinosaurs were indeed monophyletic, and he produced a phylogeny of the Saurischian dinosaurs that affirmed what Ostrom had proposed: that birds belonged within the theropod dinosaurs. In 1986, another young American student, Paul Sereno, published a phylogeny of ornithischian dinosaurs that described their relationships and affirmed their monophyly.

Today it is generally accepted that the dinosaurs are monophyletic because their first members shared the following characteristics not found in other animals:

the vomers, which are strutlike bones in the palate, are long and reach back at least to the level of the opening between the eye and the nostril;

the sacrum, which comprises the backbones that are fused into a unit with the hip girdles, includes at least three vertebrae;

the socket on the shoulder girdle where the humerus, or upper arm bone, articulates faces backward;

the deltopectoral crest of the humerus, the ridge of bone at the near end of the bone to which the muscles that raise the arm are attached, extends for at least one-third to as much as one-half the length of the humerus;

the hip socket, or acetabulum, where the ilium, ischium, and pubis meet, is not solid, but is at least partly open;

the head of the femur is set off from the shaft by a distinct neck and a rounded head;

the fibula is greatly reduced; and

the ascending process of the astragalus (a triangular flange of bone on the front of the ankle that attaches to the face of the tibia) is well developed.

These are the kinds of characteristics – trivial though some of them may seem – that scientists look at when they try to understand the relationships among groups of animals. These features were shared by the earliest dinosaurs and their descendants, and this character complex is not shared by other animals (although, for example, certain individual features may have appeared independently in other lineages of animals from time to time). This suite of features helps us to diagnose the group called Dinosauria, which we define by its component parts, Saurischia and Ornithischia.

The dinosaurs, like the Romans, had a long historical run. For 155 million years their species succeeded each other in rapid progression through the fossil record. More than 600 species have been described up to the present day, though not all of these are valid, and my colleagues Peter Dodson and Sue Dawson at the University of Pennsylvania estimated that there must have been at least 1,200 valid genera of dinosaurs between the Late Triassic and the end of the Cretaceous.

The great diversity of dinosaurs tell us that they were highly adaptable as a group and that they certainly exploited a range of ecological niches. The diversification of the Late Cretaceous hadrosaurs, or duckbilled dinosaurs, is at least broadly comparable to that of recent antelopes, given the growing but obviously incomplete record of hadrosaurs. They were diverse not only structurally but functionally; moreover, they had a lot of features that we can only interpret as sexually selective traits – those that were displayed to attract mates and repel enemies. This is why duckbilled dinosaurs, with their tremendously varied crests, can be seen as resembling African bovids, with their equally varied horns. As Darwin showed, sexual selection is an important factor in the diversification of life, and it certainly seems to have been instrumental to many dinosaurian groups.

What we call the *turnover* of species, that is, their replacement by new species as the extant become extinct, was very important for dinosaurs all through the Mesozoic Era. The 65 or so species known in the Late Jurassic all disappeared by the Early Cretaceous, from which another 90 species are known. The numbers of dinosaur families, which are loosely organized groups of related genera and species, stayed relatively stable during this time, rising from 12 to 15; but 7 of these were new.

Looking at the replacement rates of species throughout the entire span of the Cretaceous period is also important, because they can change the taxonomic picture entirely. The dinosaurian fauna of the Late Cretaceous, dominated as it was by hadrosaurs (duckbills) and ceratopsians (horned dinosaurs), had little in common with the iguanodontid-dominated Early Cretaceous. The mean duration of a species of dinosaur seems to have been about 2 or 3 million years, according to Dodson; a dinosaur genus has an average duration of about 7.7 million years. Yet despite their remarkable longevity (which seems to exceed the average figure for mammals), their great diversity, and their obvious evolutionary potential (they did, after all, give rise to birds, which are technically considered Saurischian dinosaurs in cladistic terms), the last representatives of the nonavian dinosaurs disappeared from Earth 66.4 million years ago. What can account for such an extinction?

Twenty years ago, more than 80 hypotheses had been proposed to explain this extinction. They were extremely varied, and could be classified into several categories: Some appealed to internal, biological constraints; others appealed to external factors, including abiotic, physical, terrestrial, and even extraterrestrial factors.

The *biological causes* that were invoked were diverse, and sometimes highly imaginative. Some authors imagined that the competition between species was intense, and that little by little the mammals were able to exploit the situation. Others imagined that the sex ratio – the relative numbers of males and females in the population – vacillated greatly according to the air temperature, which is known to affect the sex determination of some living reptiles in the egg. Hence, a sudden rise in temperature could have made all the embryonic dinosaurs develop into a single sex, which soon would have halted reproduction and put an end to the species. Another somewhat related hypothesis proposed that the little mammals ate all the dinosaur eggs and drove them to extinction. Another author supposed that the herbivorous dinosaurs succumbed to plant toxins that had been evolved by the newly burgeoning angiosperms (flowering plants), soon leaving the carnivorous dinosaurs with nothing to eat. Other authors have invoked epidemics, parasites, overly thin or thick eggshells (which would have dried out the eggs too quickly, or would not have allowed hatching, respectively), and many other less plausible causes (including racial senescence, genetic exhaustion, and cataracts).

The second category of hypotheses invoked *external constraints*. The dinosaurs might have perished because the rising temperatures at the end of the Cretaceous caused them to die of thirst (the version shown in Walt Disney's *Fantasia).* But other authors argued that they died from *low* temperatures! Some authors advanced the idea that changes in atmospheric pressure, or the displacement of Earth's axis of rotation, had grave consequences for the physiological tolerance of the dinosaurs. Reversals of Earth's magnetic field, volcanic eruptions, regressions of the oceans, the rise of new mountain ranges, changes in the circulation patterns of warm and cool waters as a result of tectonic activity, asteroidal impacts, the passage of comets, cosmic radiation resulting from the explosion of a supernova, and even destruction by little green men from outer space: These are only some of the causes that have been proposed and examined – some more seriously than others.

A good number of tentative explanations of these extinctions have the drawback of being able to apply conceivably only to the dinosaurs, whereas a theory with true explanatory power should also be able to explain the extinctions of other groups, whether marine or terrestrial, animal or plant, invertebrate or vertebrate, tiny or large. The oceanic extinctions included a great part of the planktonic foraminifera (the surface-floating microorganisms), ammonites and belemnites (relatives of the squid and octopus), inoceramids (large clams), rudistids (reef-building clams), marine reptiles like the plesiosaurs and mosasaurs, and flying reptiles. Such a theory also has to explain why some groups were *not* affected greatly by the extinction event, including bryozoans (small, colonial, encrusting or fan-shaped marine organisms), brachiopods (two-valved lampshells, resembling clams superficially but not at all like them), gastropods (snails), nautiloids, echinoderms, crustaceans, benthic foraminifera (those that live on the ocean bottom), fishes, lizards, snakes, sphenodontids (represented today only by the genus *Sphenodon*), champsosaurs (crocodilelike amphibious reptiles), turtles, crocodiles, birds, and many lineages of mammals (including multituberculates, marsupials, and placentals).

The extinction of the nonavian dinosaurs, and of the other components of the latest Cretaceous fauna and flora, unquestionably represents a crisis in the history of life. This crisis is not isolated, though: The history of Earth is punctuated by a series of phases of diversification of species, just as it is by crises and important decreases in its biodiversity. At the end of the Ordovician period, for example, some 435 million years ago, as well as at the end of the Devonian period, some 355 million years ago, 70 percent of the species of marine animals disappeared. The interval at or around the end of the Permian period, 250 million years ago, saw the greatest extinction in Earth's history: More than 75 percent of terrestrial vertebrate species and perhaps over 90 percent of marine species disappeared. At or around the Triassic–Jurassic boundary, about 200 million years ago, another crisis took out nearly half of the fauna. But the Cretaceous extinctions, on which so much attention has been lavished, as destructive as they were, didn't match the severity of previous mass extinctions: Only about 30 percent of the fauna disappeared 66.4 million years ago.

All through the history of life on Earth, there have been mass extinctions; but what exactly are these? A *mass extinction* is an extinction that involves species, genera, families, and other higher categories, and that takes place within a relatively brief period of time. In this interval, the rates of extinction are higher than for equivalent intervals before and afterward. This definition was given by David Archibald in 1989. The concept of a *catastrophic extinction,* however, is often confused with that of a *mass extinction.* In fact, an event can be catastrophic without causing a mass extinction.

In contrast, a *mass mortality* concerns individuals and populations; it occurs on a relatively local or regional scale, without affecting the numbers of *taxa* – species and other monophyletic groups within lineages. Contrariwise, mass extinctions can occur during brief intervals of geologic time without involving any sort of mass mortality such as a local earthquake, volcanic eruption, or other local catastrophe might cause. Indeed, the term *catastrophic extinction* should be reserved for large-scale events, such as might be caused by a worldwide system of volcanic eruptions or an asteroidal impact. Everyone who works on the Cretaceous–Tertiary boundary agrees that the end of the Cretaceous is marked by a mass extinction. A large part of the real debate is centered on the question of whether this extinction was sudden and catastrophic in all lineages (with a duration of a few thousand to about 500,000 years, according to some authors), or whether it was considerably slower and more gradual (lasting several million years).

It's interesting to consider the credible extinction hypotheses, at least those that rest on some kind of measurable or testable facts. They are relatively few, and at least four of them deserve some mention: Two appeal to sudden and catastrophic events, and two appeal to slower, more gradual processes, reflecting two different current schools of thought.

The asteroid impact. By far the most spectacular hypothesis was proposed in 1980 by a team of Californians that included the Nobel Prize–winning physicist Luis Alvarez, his son, the geologist Walter Alvarez, and their associates Frank Asaro and Helen Michel. These authors explained that the disappearance of the dinosaurs at the Cretaceous–Tertiary boundary coincided with the presence of a strong concentration of a platinum-group metal called iridium. This concentration was so strong, they said, that it must have had an extraterrestrial origin. The iridium signal had been found in marine-bottom sediments deposited at Gubbio in Italy, then at Stevns Klint in Denmark, and then at many other locations, including the deep-water deposits at Zumaya in Spain. The team concluded that this high concentration of iridium could only have come from the impact of a huge asteroid, 6–15 km in diameter (4–10 mi.). The presence of shocked quartz grains and of minerals rich in nickel in the same deposits, which were studied by the Frenchman Robert Rocchia, confirmed the existence of a collision so great that it would have created shock waves, giant tsunamis, extreme tides, widespread fires, and, in the long run, a darkening of the skies, severe acid rains, and an exaggerated greenhouse effect that resulted in a broiling increase

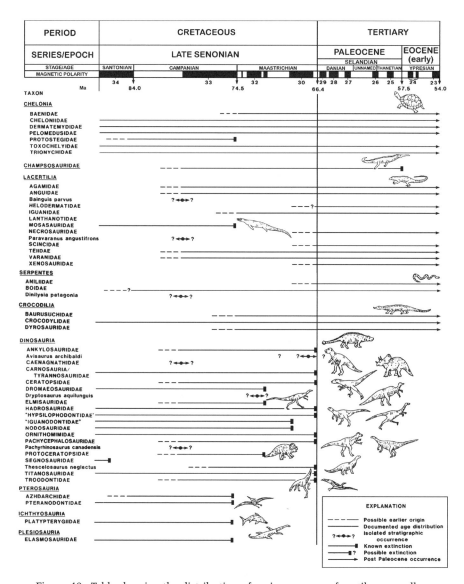

Figure 40. Table showing the distribution of various groups of reptiles, as well as some isolated species, at the end of the Cretaceous and the beginning of the Tertiary. The groups that became extinct at the very end of the Cretaceous were represented by only one or two species. These groups were strongly reduced in diversity at the end of the Campanian and throughout the Maastrichtian. (After Sullivan 1987, modified by F. Pilard)

and then a chilling decrease in temperature. All these events would have had catastrophic effects on the fauna and flora of Earth. For several years the search was on for the impact crater of such an asteroid, which certainly would have had respectable dimensions much larger than the asteroid itself. A number of sites were proposed and examined, and today the proponents of the hypothesis appear to agree that the culprit is the Chicxulub crater, off the Yucatán Peninsula of Central America, in the western Atlantic Ocean.

The volcanoes of the Deccan Traps. In 1983, another American team, composed of the American geologists Charles Officer and Frank Drake, and joined in 1986 by French teams under the direction of Vincent Courtillot, concluded that an episode of severe volcanism coincided with the Cretaceous–Tertiary boundary. Intense lava flows, called the *traps* of the Deccan Plateau of India, threw up enormous quantities of gases and sulfuric acid over a period of 500,000 years. In fact, iridium really is a by-product of volcanic eruptions. (There is no iridium in the Deccan Traps themselves, but it is abundant in the particles ejected by the Kilauea volcano and the Piton de la Fournaise on Reunion Island, for example.) Quartz and its related crystalline forms, decorated with microscopic bubbles, show that they were formed at very high temperatures, which is consistent with a volcanic explosion. The consequences of such eruptions would have been very similar to those of a meteorite impact: darkness, complete or nearly complete cessation of photosynthesis, then intense cold accompanied by abundant acid rain and catastrophic extinction of the fauna and flora.

Marine regression and globally lower temperatures. This hypothesis was advanced in 1964 by Léonard Ginsburg, a paleontologist at our Museum of Natural History in Paris, and he presented updated evidence for it in 1984; it has since been taken up by other authors. Ginsburg argued that the disappearance of the dinosaurs coincided with a regression of the seas at the end of the Cretaceous, which correlated with the development of glaciers. The extension of newly emerged land surfaces increased the continentalization of climates (during which time the winters would become colder and the summers warmer, as they are in Mongolia). The resulting lowered temperatures would have affected the animals that were sensitive to cold, and caused a drastic reduction in the habitable surface area available to marine organisms. Today we can measure the temperatures of the past, which we call *paleotemperatures,* by assessing the relative abundance of oxygen isotopes (^{16}O and ^{18}O) in calcium carbonate crystals in the shells of marine organisms. This relative abundance reflects the abundance present in the seawater when the organism's shell was formed, and it varies according to the water temperature. The paleotemperatures obtained by Larry Frakes, an Australian researcher, showed in 1979 that there had been a progressive and gradual decrease in surface water temperatures through the Late Cretaceous.

Competition among species. A number of paleontologists – such as the Americans Leigh Van Valen and Robert Sloan in 1977, then another Californian team headed by Bill Clemens beginning in 1981, and their colleagues from the East Coast, Max Hecht and Antoni Hoffman in 1986 – favored explanations that invoked gradual and progressive changes in the environment that had mostly to do with competition among species, following the classic neo-Darwinian paradigm. There were a number of cases in which this process applied perfectly. For example, the evolution of the faunas of North and South America during the Cenozoic or Tertiary Era, and the disappearance of many of these species, depended at the same time on biogeography (the separation between the nothern and southern continents) and competition between species. When the Isthmus of Panama was established, the North American mammals spread through much of Central and South America, gradually supplanting and eliminating the native mammalian fauna, including Edentates such as *Megatherium* and *Glyptodon.* There seems to be no obvious reason why Mesozoic mammals, which were more numerous, more active, nocturnal, and capable of hiding in burrows and hibernating, couldn't have supplanted the dinosaurs at the end of the Cretaceous when climatic conditions became unfavorable.

A model complementary to competition among species concerns the general deterioration of the biotic environment as climate caused plant associations to change, thereby affecting the animal communities that depend on them for food. This model might be applied to the well-documented example that Tassy and Saunders provide of the disappearance of the North American mastodon some 10,000 years ago. This mastodon inhabited the continent for two and a half million years. During that time there were a considerable number of climatic oscillations that caused the climate to vary between warm temperate and harsh glacial phases. Eventually these vacillations led to the breakup of the vast forests, which fragmented the last mastodon populations into islands; when these islands disappeared, the mastodons disappeared along with them. An intense ecological stress, lasting only three to four thousand years, was enough to account for the disappearance of this species, without resort to any special sudden catastrophe. A similar ecological disturbance that resulted in the fragmentation of habitats into islands surrounded by water occurred during the latest Cretaceous of central Europe, southern France, and Spain, and it is correlated with a sharp reduction in size of several lineages of dinosaurs. A similar rapid trend in size reduction took place on many newly created Mediterranean islands during the rise in sea level of the early Pleistocene; this stimulated the rather incongruous evolution of giant dormice and swans, along with repeated "dwarf" lineages of mastodons, mammoths, and even hippos.

These are very brief summaries of the four main hypotheses – at least the serious ones – that currently vie for the attention of the scientific community. If we took a poll among researchers, it might be fair to say that statistically the majority of astrophysicists, geophysicists, and volcanologists tend to favor the *catastrophist* hypotheses, whereas most paleontologists tend to favor the *gradualist* hypotheses. A small number of researchers have no opinion, and some have just completely had it up to here with the whole thing.

To advance hypotheses is the business of the scientific community. But how can we test them? The main evidence we have is buried in the ground; that is, the response of organisms to the end of the Cretaceous. This response is recorded in the geologic horizons of the Cretaceous–Tertiary transition, or *K/T boundary:* Here, *K* stands for *Kreide,* German for chalk, because the classic Late Cretaceous European marine sections are known for their limestone or chalk cliffs made almost entirely of the skeletons of microorganisms. (Think of the White Cliffs of Dover, and if you have the chance, look for Thomas Henry Huxley's famous essay, "On a Piece of Chalk.") *T* stands for Tertiary, the "third" major interval in the history of Earth (the "Primary" and "Secondary," however, don't exactly correspond to the Paleozoic and Mesozoic Eras). So the K/T boundary is the interval that spans these great geological entities.

It turns out that the response of the biosphere in general to the K/T boundary event was extremely variable among groups, and so there is no simple answer to the riddle. Each hypothesis has problems in accounting for all the facts provided by the archives of Earth: the fossil floras and faunas.

In the oceans, the ammonites and belemnites saw the numbers of their species decline throughout the Cretaceous, particularly between the Albian and Aptian stages, then between the Campanian and Maastrichtian stages – that is, essentially through the whole duration of the Late Cretaceous (some 30 million years). So these groups were close to extinction long before the end of the Cretaceous, and nearly all the ammonites disappeared more than 100,000 years before the infamous Cretaceous–Tertiary boundary. In contrast, the number of bryozoan families increased from the Cenomanian to the Campanian (that is, through nearly the entire Late Cretaceous), stabilizing through the Maastrichtian (the last stage of the Cretaceous) and into the Paleocene (the start of the Tertiary), and finally increasing again thereafter. Planktonic foraminifera took a very hard hit, however: 36 species were present at the end of the Cretaceous, but only one is found at the start of the Tertiary.

On the continents, the floras, especially those composed mostly of flowering plants (angiosperms), developed all through the Cretaceous. Some geologic sections show a sharp decline in pollen abundance at the end of the Cretaceous, compared to spores of ferns and related plants, but the various families and lineages of flowering plants don't seem to have suffered any kind of mass extinction during the same period: They reappear later in the geologic record.

One of our American colleagues, Bob Sullivan, did a study in 1987 in which he looked at the diversity of reptiles during the last interval of the Cretaceous and the first interval of the Tertiary (from 87 to 54 million years ago). He found out some interesting things that need to be taken into account. His detailed examination of all these fossil species in their precise stratigraphic positions showed convincingly that there was a gradual process of extinction of species and higher taxa (genera, families, etc.) through the K/T transition. This process is not compatible with any hypothesis that calls for the simultaneous extinction of groups in a relatively short period of time. To the contrary, a minor wave of extinctions began at the end of the Campanian stage (less than 74.5 million years ago), eight million years before the K/T boundary. The victims included lineages of turtles (the Protostegidae), lacertilians (the mosasaurs, including those from Maastricht!), flying reptiles (the Pteranodontidae, with a long beak, skull crest, and reduced tail), the last ichthyosaurs (the dolphin-shaped Platypterygiidae), and the last plesiosaurs (the extremely long-necked Elasmosauridae). Then, during the Maastrichtian, two waves of extinction followed: Five million and then three million years before the end of the Cretaceous, five dinosaur lineages disappeared. There were only ten dinosaur lineages left as they neared the K/T boundary, and these represented only 12–14 species that were among the last survivors. As Bill Clemens's crews have shown in Montana, in the last dozens of meters before the K/T boundary there were only two species of dinosaurs left – *Tyrannosaurus* and *Triceratops* – and neither they nor any other land animals are piled up as carcasses at the boundary; in fact, their last remains are well beneath it. These facts were ironically underlined by a question put by Mike Benton: Does the loss of 15 species in three million years constitute a catastrophic extinction?

It seems more logical to suppose that at the end of the Cretaceous there was a series of ecological stresses that led to the progressive disappearance of a part, and only a part, of the marine and terrestrial faunas. But catastrophic approaches need to be very subtly phrased in order to take into account the diversity of responses by various organisms to the events at the end of the Cretaceous.

Recent work has reinforced the views of the gradualistic proponents. A more detailed look at the brutal decline of planktonic foraminifera showed Gerta Keller, a micropaleontologist at Princeton University, that the responses of different species within this group at the K/T boundary were not even identical between the Northern and Southern Hemispheres: In one case, the decrease in species was progressive and in the other it was sudden, although the same species were involved.

In June 1991, three Australian researchers, Malcolm Wallace, Reid Keays, and Victor Gostin, showed that the concentration of iridium at certain geologic levels could be caused by something completely different from meteoric impact or a volcanic ashfall. In two geologic levels from the Paleozoic Era and one from the Mesozoic, the concentration of iridium seems to have been produced as a result of a regression in sea level that led to a condensation of elements that are normally more dispersed in the vast oceanic masses. Platinum-group metals, including iridium, could thus have been concentrated on the ocean bottoms. This enrichment could have been accentuated by the presence of *stromatolites*, mats of blue algae that are common from very early in the fossil record and survive today along large stretches of the Australian coast. These mats of blue algae can use microbiological processes to concentrate rare elements dissolved in sea water as they construct their massive, mound-like reefs. So in some cases an iridium concentration can simply be the result of a biological process that is produced during a regression of the oceans.

Finally, Larry Frakes, an expert on paleotemperatures, published a very interesting article in 1994 in the *Comptes rendus de l'Académie des sci-*

ences in which he confirmed the importance of gradual cooling at the end of the Cretaceous. Frakes and his collaborators observed a telling correlation: The Maastrichtian was one of the coolest intervals, whereas the late Paleocene, well into the Tertiary, began a much warmer interval.

Beyond these diverse attempts to explain the phenomenon of mass extinctions, however, we really have to ask a whole series of epistemological questions, to bring the perspective of the historian of science to ideas regarding the notion of extinction.

There are currently a lot of difficulties that surround the interpretation of great changes in evolution, the discontinuities in the history of Earth and its life. But these problems are inherent to geology, paleontology, and biology; indeed, they have been encountered ever since the beginnings of these sciences. In order to explain the turnovers and the relay races that we see in the historical patterns that constitute the evolution of species, scientists have tried to tease out the parts caused by external physical constraints from those due to internal biological constraints. Even though we know that evolution is the result of an interaction between organisms and their environments, the importance accorded to one or the other class of factors has varied through time. At the beginning of the nineteenth century, in the middle of the post-Revolutionary period in France, Cuvier declared that organisms had been subjected to what he called "revolutions of the globe," large-scale, catastrophic changes at intervals in Earth's history: Living organisms were destroyed, perishing suddenly. His disciple, Alcide d'Orbigny, concluded that 27 such revolutions had punctuated geologic history, each one corresponding to a clean break in the succession of the archives of Earth. The French geologist Élie de Beaumont then made each extinction coincide with the uplift of a mountain chain.

In Cuvier's time, his colleague and friend the geologist Constant Prévost, in France, and then Charles Lyell (who wrote the famous *Principles of Geology*), writing in an England that believed strongly in the progress of humans and society, proposed a more gradualistic approach, based on actualistic (present-day) processes and *uniformitarianism* (the idea that the laws, processes, and even sometimes rates of Nature were the same in the past as they are now). They were opposed to Cuvier's catastrophic vision, which postulated that certain past events could be explained only by special and severe causes. Following Prévost and Lyell, Charles Darwin wrote in *On the Origin of Species* that the extinction of old forms and the production of new and improved forms are intimately connected together. The old notion of all the inhabitants of the earth having been swept away at successive periods by catastrophes, is very generally given up. . . . It is most difficult always to remember that the increase of every living being is constantly being checked by unperceived injurious agencies; and that these same unperceived agencies are amply sufficient to cause rarity, and finally extinction.

Until the mid-nineteenth century, the living world was viewed as a system that was externally regulated. In Darwin's theory of the evolution of species by means of natural selection, the forms of living beings, their properties, and their characteristics were regulated by a vast system that included the Earth and all the objects on it. This is the model that was accepted and is still followed today by Darwin's intellectual descendants.

With the research of the Alvarez group in the 1980s, a new approach to extinctions developed, but this time it was based on neocatastrophism. If extraterrestrial extinctions are more frequent, more rapid, and (most important) more sudden and wide-ranging in their effects than we'd suspected, then the processes of *microevolution*, the small transformations of genes and living organisms by mutations, are inadequate to explain the evolution of lineages and communities of organisms. With these kinds of effect, the adaptations of organisms to their local environments would lose virtually all significance in the grander scheme of things. Instead, a previously unsuspected selective regime, adapted to catastrophes, might assume primary importance on the large scale. Paleontologists who have analyzed this possibility, such as David Jablonski and Stephen Jay Gould, have even suggested that there might be a general theory of mass extinction that would supplant the neo-Darwinist perspective in the consideration of macroevolutionary change.

In the final analysis, we must ask whether the distribution of living organisms over Earth's surface really is the result of known, everyday processes of evolution, of the play of possibilities, of interaction between what was and what will be, between the conservative and the revolutionary, between the fidelity of reproduction and the novelty of variation (as François Jacob asked in his book *The Logic of Life*); or do we find that the results of Cuvier's revolutions of the globe, the survival of the luckiest (as opposed to the best adapted), introduce completely new rules – if terrestrial events (like meteorite showers) or events tied to the internal dynamics of Earth's crust (like volcanic eruptions) play a role that is ultimately much more important than we thought?

The proponents of end-Cretaceous volcanism have extended their hypothesis to cover all the great crises in the history of life: The principal intervals of extinction coincide with the great phases of volcanic eruptions. This approach (which recalls Élie de Beaumont's) underscores the feeling that the history of life has a lot to do with drawing lots. On the other side, the proponents of neo-Darwinism reply that the high frequency and periodicity of mass extinctions is doubtful. These extinctions are based only on statistics insofar as they define *extinction* in an unusual and idiosyncratic way, one that doesn't call for a large interval of time. There is no sure and incontestable criterion that allows us to distinguish mass extinctions as classes of distinct phenomena, and there are few reasons to think that they pose a major challenge to the conceptual utility of neo-Darwinism. For these authors, more than ever, neo-Darwinism is the only explanation for the evolution of species: Nothing allows us to be sure that the processes we observe, whether in periods of great expansion of living organisms or in times of their mass extinction, reflect anything other than the results of microevolutionary forces operating on a great number of biological species, under the influence of normal physical and biological conditions.

Discontinuities in the fossil record are (ironically) a stroke of good fortune for geologists, paleontologists, and also historians: Without discontinuities, there can be no chronology. They are certainly obstacles to the idea that continuous processes comprise the origin and history of Earth, the lithosphere, and life itself. This problem is really not unique to the earth sciences; it was treated elegantly by the Dutch epistemologist Hooykaas in a work called *Natural Law and Divine Miracle: A Historical-critical Study of the Principle of Uniformity in Geology, Biology, and Theology.* Hooykaas showed that after 1840 "we find ourselves in the presence of three different understandings: rigorous uniformitarianism to explain the history of the Earth and the history of life (as Lyell conceived it); catastrophism in geology and progressivism in biology (the view of the English geologist Sedgwick), and uniformitarianism in geology and gradual progression in the organic world (as Darwin conceived it)." Today, we could add a new kind of understanding: neocatastrophism in geology and punctuated equilibria in the evolution of species (as Eldredge and Gould conceive it). The hypothesis of *punctuated equilibria* implies that evolution proceeds at different rates: There are long periods of stasis, that is, the cessation or extreme slowing down of evolutionary processes and changes in patterns, punctuated by short periods when evolution becomes very rapid. We could conclude from this that there are many ways to interpret the tempo and mode of geological and biological events. Some authors are proponents of the "strongly continuous" view and others are of the "strongly discontinuous" persuasion, and there are many versions of a "continuous–discontinuous" mixture involving both geological and biological processes.

To Hooykaas, moreover, it seemed that among the savants of the nineteenth century there was a mutual influence, explicit or tacit, between their visions of the world and their scientific interpretations. It would be amazing if the same were not true of scientists today. In 1844 Robert Chambers, Darwin's contemporary and a proponent (as was Darwin) of uniformitarianism in geology and progressivism in biology, violently attacked the catastrophic view, based as it was on the text of Genesis, in his anonymous book *Vestiges of Creation*. The questions of "continuous" vs. "discontinuous" were not limited to biology and geology, as he well knew. They also applied to the domains of metaphysics and history.

The extinction of dinosaurs at the end of the Cretaceous is a fascinating event, but it's inseparable from the patterns of disappearance and survival of all the other forms of life that populated Earth at the same time.

Analyses of the causes of this event must be directed toward constructing a theory of crises that can help us understand *all* the great changes that have punctuated Earth's history. The eventual role of this theory will be to explain the survival, decline, or revival of the success of individuals and the diversification of species and lineages. The dinosaurs are only one example among thousands that must be understood.

After having dominated Earth for 155 million years, and after a brief decline of a few million years, the last of these famous dinosaurs (excepting birds) finally disappeared from our planet 66.4 million years ago – not from the action of a single cause, sudden and brief, but more likely from a multitude of complex causes, stretched over a long interval.

After leaving their traces at the four corners of the Earth, the dinosaurs, those surprising travelers from the Mesozoic, have returned today, thanks to the bone hunters, to leave their mark on the world once again – a world dominated by a single species, highly prolific, highly adaptable, and perhaps the least bit irresponsible – namely, our own. May the fate of the dinosaurs be a lesson in reflection and meditation to the hominids that they continue to fascinate.

AFTERWORD

PHILIPPE TAQUET
AND
KEVIN PADIAN

IN THE FEW YEARS between the original publication of this book in France and its appearance in the translated form that you hold, a great deal has happened in the world of the dinosaurs – even though all the dinosaurs but the birds have been extinct for some 65 million years. These pages have narrated a travelogue through time and space, one paleontologist's journey through 30 years of his science. In these final pages we would like to update our travelogue, showing the reader some of the most recent adventures and discoveries of our bone-hunting brethren.

This book's story began in the north of Africa, in the exotic countries of Niger and Morocco. Here, the researches of geologists and paleontologists such as de Lapparent and Monbaron uncovered the beginnings of what we know was a rich history in places historically viewed as little more than desert and arid mountain range. The excavations at Gadoufaoua, Wawmda, and other places carried on the tradition of de Lapparent and explorers such as Theodore Monod, who loved the desert like all wild places, areas of solace and discovery.

New expeditions to northern Africa have expanded on earlier discoveries and brought us new insights. The Triassic beds of the High Atlas of Morocco were explored by crews from the Museum in Paris during the 1960s and more recently by Farish Jenkins and his colleagues from Harvard University in the 1990s. De Lapparent's work was taken up by Paul Sereno and his colleagues from the University of Chicago, who found additional sauropod dinosaur material as well as new material of the giant theropod *Carcharodontosaurus* and a new basal coelurosaur that they named *Deltadromeus*. These crews had also explored the deserts of Niger several years before and returned with the skeleton of another vicious predator, *Afrovenator.* These finds and others built on earlier conclusions that Africa had been connected to other continents, such as Europe and South America, well into the Early Cretaceous. Some Late Cretaceous dinosaurs also apparently migrated freely between Africa and other continents. Even more recently, for the first time in many decades, Madagascar has been opened politically to travel and exploration. It was not long before David Krause, Cathy Forster, and their colleagues from the State University of New York at Stony Brook and other institutions began to bring back Late Cretaceous dinosaurs and birds that are opening a whole new window on that part of the world.

During the 1960s and 1970s, the evidence from dinosaurs, crocodiles, fishes, and crustaceans was leading Taquet, Buffetaut, and other colleagues to realize that there had been a connection between Africa and South America until well into the Early Cretaceous. At the same time, other research in South America was bringing to light the first remains of small bipedal archosaurs that turned out to be the closest relatives to dinosaurs yet known.

The African–American connection was first established between Niger and Gabon on one side, and Brazil on the other. Brazil has given us a few very interesting but poorly known dinosaurs; on the other hand, Argentina, from the badlands of La Plata and Ischigualasto to the southern reaches of Neuquen and Patagonia, has changed forever how we think about the Age of Dinosaurs, from its beginnings to its end. Some of the first surprises came from field parties from Harvard University in the 1950s and 1960s, led by A. S. Romer and Jim Jensen, in collaboration with the great Argentinian vertebrate paleontologist José F. Bonaparte. From Ischigualasto they brought back late Middle and early Late Triassic reptiles in stubborn nodules of rock that contained a hash of small skeletons and skeletal parts. It took many years for these materials to be prepared and sorted out, but they turned out to represent a whole diversity of small archosaurs never before seen. Some of these animals were crocodile relatives, but a few were clearly closer to true dinosaurs, as their long limbs, hingelike ankles, and upright posture showed. These animals were no larger than rabbits, and so Romer named them *Lagosuchus* ("rabbit crocodile") and *Lagerpeton* ("rabbit reptile"). Bonaparte produced a seminally important monograph on *Lagosuchus* in 1975, showing that in many respects it straddled the borders between typical archosaurian reptiles (then called "thecodontians") and true dinosaurs. In fact, for the better part of five decades, José Bonaparte has been the leading light in South American vertebrate paleontology, exploring its Mesozoic deposits for all manner of ancient dinosaurs and their relatives, exotic reptiles poorly known elsewhere in the world, and the first known South American mammal relatives. His former students are now dispersed throughout Argentina and the world, carrying on new research and collaborating, as their professor always has done, with colleagues all over the globe. As a result, our knowledge and the importance of the Argentinian fossil record continues to grow by leaps and bounds.

In the late 1980s and early 1990s, Paul Sereno and José Bonaparte returned to Ischigualasto and found that the fossil beds were by no means exhausted. They collected a variety of Late Triassic reptiles and distant mammal relatives. In succeeding years Sereno continued to work with Bonaparte's students and field colleagues, such as Andrea Arcucci, Fernando Novas, and Oscar Alcobar. These colleagues revised and updated a substantial amount of earlier work, but they also discovered new skeletons of two key animals: *Herrerasaurus*, long thought to be some sort of primitive dinosaur (possibly a saurischian), and *Eoraptor*, an entirely new form smaller than *Herrerasaurus* and somewhat more basal. Recent descriptions have portrayed these animals as basal theropod dinosaurs, and perhaps they are, although they may also be the closest animals we know to dinosaurs without being true dinosaurs themselves. The full analyses of these animals have not yet been set before the scientific community for evaluation, but the results promise to be most interesting.

Meanwhile, in southern Argentina, Rodolfo Coria, Luis Salgado, and José Calvo have been uncovering fantastic new dinosaurs from the medial and Late Cretaceous. One example is the short-faced, bull-horned large theropod *Carnotaurus*, which Bonaparte, Novas, and Coria brought to light some years ago. It appears to be a relict of a basal theropod lineage that may have split from other theropods before the Late Jurassic, at least. Coria and Salgado have described several other dinosaurian remains, including ornithischians and sauropods, but one of their most celebrated finds must surely be the giant carnivorous dinosaur *Giganotosaurus* from Patagonia – which, along with *Carcharodontosaurus* from northern Africa, rivals and perhaps exceeds *Tyrannosaurus* in size. Of course, these animals are not identically shaped, so it is not easy to tell which is the longest, the heaviest, the tallest, which is graced with the largest skull or sharpest teeth, and so on. Neither can we tell whether these specimens are the largest members of their populations. What is particularly interesting is that *Carcharodontosaurus* and *Giganotosaurus* appear to be closely related, implying once again that a connection between their southern continents persisted well into the Cretaceous.

It seems a little strange that ornithischian dinosaurs are not better known from the southern continents in the Late Jurassic and Cretaceous. Certainly they were present; in fact, the oldest and most basal ornithischian remains are known from South America (Bonaparte's *Pisanosaurus*, from the Late Triassic) and South Africa (the Early Jurassic *Lesothosaurus* and *Heterodontosaurus*, for example). Moreover, they obviously flourished in North America, Europe, and Asia. In Africa, only *Ouranosaurus* is well known from the Early Cretaceous, and we have only partial remains of ornithischians from Southeast Asia, as noted in Chapter 7.

Work in Laos has continued to the present, and investigators now agree that the Laotian fossil-bearing beds are most probably of Early Cretaceous age, considerably older than J.-H. Hoffet estimated in the 1930s; the dinosaurian remains are probably not those of Late Cretaceous–type titanosaurs and hadrosaurs, as he thought, but of more basal taxa characteristic of the Early Cretaceous. Still, there is no denying Hoffet's astonishing insights, given the information available to him at the time in this remote outpost. Meanwhile, in neighboring Thailand, across the Mekong River, our French colleague Eric Buffetaut has collaborated with a number of Thai workers to bring to light dinosaurian remains ranging in time from the Late Triassic to the Early Cretaceous – apparently contemporaneous with the Laotian forms. Most of the Thai dinosaurs, like the Laotian ones, are fragmentary, represented by teeth and isolated bones; but there are jaw and teeth remains of the basal ceratopsian *Psittacosaurus*, and well-represented skeletons of a new sauropod dinosaur, *Phuwiangosaurus*, both from the Early Cretaceous. There are also some unusual theropod remains and discoveries of trackways, suggesting a tantalizing fossil record that must eventually yield more treasures and surprises.

Psittacosaurus is mainly known from Mongolia, of course, and it is notable because it represents the most basal member of a group that radiated through Asia (*Protoceratops* and its relatives) and later into North America (more protoceratopsids, but especially the great ceratopsid radiation that included the chasmosaurs and centrosaurs, including *Tricera-*

tops, Torosaurus, Chasmosaurus, Pentaceratops, and many other forms). We encountered *Protoceratops* at the Flaming Cliffs in Mongolia, right where Roy Chapman Andrews and his colleagues had found them in the 1920s. It is still perhaps the most common dinosaur in those Mongolian beds. However, new research is beginning to tell a whole new tale of these deposits and their denizens.

It will be remembered that we ran into our colleagues from the American Museum of Natural History (AMNH) in the vast badlands of Mongolia in 1991. Since then, these intrepid crews have returned year after year, integrating new discoveries with the information from specimens collected by the Central Asiatic Expeditions in the 1920s, as well as by the Soviet–Mongolian expeditions in more recent decades; yet the new discoveries seem to be as spectacular as the earlier ones. Messrs. Norell, Novacek, Clark, McKenna, Dashzeveg, and their colleagues have found new and better material of the strange, toothless, crested theropod *Oviraptor,* which Henry Fairfield Osborn named from remains brought back by the Andrews–Granger team in the 1920s.

The name *Oviraptor* means "egg-stealer," and the specific name *philoceratops* means "fond of the horn-faced." Osborn and his colleagues had concluded that the toothless, sharp-beaked predator was snatching the eggs of the ubiquitous *Protoceratops,* which historically has been portrayed, in countless museum exhibits and picture books, in the company of its eggs and babies. However, new finds revealed *Oviraptor* entombed in a sand dune while it was sitting on its nest. Its arms and legs are drawn up around the eggs – but the eggs are those traditionally attributed to *Protoceratops*! There can be no doubt now that these are *Oviraptor* eggs, because the bones of embryos have now been found within some of the eggs. (More surprisingly, there are associated finds of other eggs belonging to dromaeosaurids. Were there other nesting species in the area, or did the dromaeosaurs slip the odd egg into a strange nest, cuckoolike?) So it appears that the "egg-stealer" has been granted a pardon, and instead can be held up as a model parent, sitting on the nest while, it is thought, a sandstorm buried all the animals in the area. New research on the sedimentology of these deposits, however, suggests that something like mudslides may have suddenly entombed these creatures, as the dunes of sand and clay collapsed precipitously.

In addition to these spectacular finds, the AMNH crews have discovered new mammals with strikingly unusual features, a vast array of fossil lizards, and a number of birds, including the bizarre Late Cretaceous form *Mononykus*. This animal is obviously close to birds, and several features place it slightly closer to living birds than is the famous *Archaeopteryx*. However, *Mononykus* is not called "bizarre" for trivial reasons: Its forelimbs are very short – too short even to reach its mouth – and all the fingers except the first have been lost or reduced to insignificance. Yet *Mononykus* has a powerful crest on its humerus, or upper arm bone, for rotating the arm; it also has a bony process on its elbow that is about as long as the rest of the forearm itself, and obviously had some sort of function in extending the middle part of the arm. Perhaps as strange as these features is the blocky wrist and the single, attenuated, spikelike finger. What was it doing with this strange apparatus? Certainly it did not fly, though it most probably descended from flying ancestors. The prominent crests and processes for muscle attachment on the forelimb have suggested to some a molelike digging; but the forelimb is so short that one wonders whether any such function was still possible in *Mononykus,* though perhaps its ancestors had had longer and more functional arms. Was there perhaps some sort of modified feathering, not preserved in specimens collected to date, that would help to answer these questions of its function and ecology? So far, *Mononykus* remains largely a mystery in those terms; but its relatives, which belong to the lineage named Alvarezsauridae, are known from the Early Cretaceous of South America and the Late Cretaceous of North America – suggesting a long history and considerable time for modification and ecological diversification to take place.

It would take an entire book to describe the discoveries of dinosaurs that have been made in China since the 1960s; in fact, several very good ones have been written. The "godfather" of Chinese Mesozoic vertebrate paleontology, Yang Zhongjian (C. C. Young), lay the groundwork for much of the explorations of more recent times; his productivity ensured that the science of vertebrate paleontology would continue in China, even through lean political and economic times. When China was reopened to the West, foreigners were treated firsthand to a host of accomplishments by their Chinese colleagues, many of whom they had not seen in decades. Of these colleagues, special mention must be made of Dong Zhiming insofar as dinosaurs are concerned. Dong and his colleagues and coworkers amassed legendary collections from Mesozoic beds throughout China: hadrosaurs and pterosaurs from Xinjiang, tons of sauropods and other dinosaurs from Zigong, Early Jurassic sauropodomorphs and theropods from Lufeng. However, a particular effort was made to explore the Middle Jurassic beds of China, particularly in Yunnan, Guizhou, Sichuan, and Zigong, because this time interval has historically been very poorly known throughout the world. Their efforts were rewarded by the discoveries of Middle Jurassic sauropods, theropods, and stegosaurs, ornithopods and pterosaurs, as well as amphibians, fishes, and even a plesiosaur – some 40 excavated tons representing eight thousand bones – all in all, the richest Middle Jurassic record in the world, mostly from the Dashanpu Dinosaur Quarry near Zigong City. The Chinese, in opening their doors to Westerners, also reaped the benefits of international collaboration. One of the first fruits of these labors was the Sino–Canadian Dinosaur Project, undertaken with Philip Currie, Dale Russell, and colleagues from Alberta's Royal Tyrrell Museum of Palaeontology in the late 1980s. This work resulted in the discovery of 11 new species of dinosaurs, including eight new genera. Collaboration continues with teams from London's Natural History Museum, Texas Tech University, Harvard University, and other institutions.

It is a common axiom in paleontology that "the more you look, the more you find," and recent expeditions all over the world continually prove its truth; yet nowhere is it more true, at times, than in your own backyard. Although European dinosaur finds have generally not been as spectacular as those from other continents, they are often no less important. For example, it was in Europe that the first records of Middle Jurassic stegosaurs were recorded. These earliest known stegosaurs were represented only by scrappy remains and not discovered until the 1980s; but almost immediately thereafter, better material of Middle Jurassic stegosaurs surfaced in China. A partial but very important early ankylosaur is also known from the Middle Jurassic of Europe, although the group did not really diversify until the Late Cretaceous. In addition, both the Isle of Wight in England and the Transylvanian region of Romania have now produced a variety of Early Cretaceous dinosaurs, some new and some known or with close relatives elsewhere in the world. Just recently the first discovery of eggs and embryos of theropod dinosaurs in the Late Jurassic of Portugal was announced, providing potentially new information about dinosaurian development and behavior. New finds of European dinosaurian remains near the Cretaceous–Tertiary boundary may suggest a pattern of extinction rather different from that seen in the Western Interior of the United States and Canada.

For now, the importance of many of these new discoveries is still to be fully realized. Good science, like good wine, takes time to develop and mature. Any summary work, like this one, is no more than a pause in the continual rush to discovery. For readers interested in more details of recent discoveries and discussions, we suggest the following books, with the hope that dinosaurs will always leave a vivid impression on your scientific imagination.

Currie, Philip J., and Kevin Padian (eds.). 1997. *Encyclopedia of Dinosaurs.* San Diego: Academic Press.

Farlow, James O., and Michael K. Brett-Surman. 1997. *The Complete Dinosaur.* Bloomington and Indianapolis: University of Indiana Press.

Glut, Donald F. 1997. *Dinosaurs: The Encyclopedia.* New York: McFarland & Co.

BIBLIOGRAPHY

THE FOLLOWING LIST of publications does not pretend to scratch the surface of the bibliography that could be produced related to the subjects discussed in this book. My goal was to include both technical and popular references, including those that I have used in my work, as well as some that I have published myself and with my colleagues. I hope that English-speaking readers will find some sources that are new to them but no less important and interesting than those already familiar.

1. GADOUFAOUA: IN THE SANDS OF THE TÉNÉRÉ

Doyle, Arthur Conan. 1912. *The Lost World.* New York: Review of Reviews.

Faure, Hughes. 1966. "Reconnaissance géologique des formations sédimentaires postpaléozoïques du Niger oriental," *Mémoires du Bureau de recherches géologiques et minières,* no. 47: 1–630.

Lapparent, Albert-Félix de. 1960. "Les Dinosauriens du continental intercalaire du Sahara central," *Mémoires de la Société géologique de France,* 39(58A): 1–56.

Monod, Théodore. 1937. *Méharées – Explorations au vrai Sahara.* Paris: Éd. Je Sers.

Taquet, Philippe. 1966. "Mission dinosaures," *Sciences et avenir,* no. 237: 762–7.
   1967. "Découvertes paleontologiques recentes dans le nord du Niger," *Problèmes actuels de paléontologie – Évolution des vertébrés* (colloque international du CNRS), 415–18.
   1972. "À la recherche des dinosaures du Niger," *Le Courrier du CNRS,* 3: 33–6.

Teilhard de Chardin, Pierre. 1956. *Lettres de voyage, 1923–1955.* Paris: Grasset.

2. THE *OURANOSAURUS;* OR, HOW TO BRING A DINOSAUR BACK TO LIFE

Bultyinck, Pierre. 1989. *Bernissart et les Iguanodons.* Bruxelles: Éd. Institut royal des sciences naturelles de Belgique.

Cuvier, Georges. 1800–5. *Leçons d'anatomie comparée de G. Cuvier recueillies et publiées sous ses yeux par L. Duvernoy, etc.*, 5 vols.

    1812. *Discours préliminaire à l'ouvrage: Recherches sur les ossemens fossiles de quadrupèdes ou l'on rétablit les caractères de plusieurs espèces d'animaux que les révolutions du globe paraissent avoir détruites.* Paris: Deterville, 4 vols., 1821–4; repr. Paris: Christian Bourgois, 1985, with a preface by Herbert Thomas and afterword by Goulven Laurent.

Ellenberger, François. 1988. *Histoire de la géologie*, vol. 1. Paris: Lavoisier, Petite Collection d'histoire des sciences, Technique et documentation.

Gardom, Tim, and Angela Milner. 1993. *The Natural History Museum Book of Dinosaurs.* London: Virgin Books.

Gohau, Gabriel. 1987. *Histoire de la géologie.* Paris: La Découverte.

Gould, Stephen Jay. 1993. *Bully for Brontosaurus.* New York: W. W. Norton.

Horner, John. 1992. *Cranial Morphology of* Prosaurolophus (Ornithischia: Hadrosauridae) *with Descriptions of Two New Hadrosaurid Species and an Evaluation of Hadrosaurid Phylogenetic Relationships.* Bozeman, Montana: Museum of the Rockies, Occasional Paper no. 2.

Matile, Loïc, Pascal Tassy, and Daniel Goujet. 1987. "Introduction à la systématique zoologique (Concepts, Principes, Méthodes)," *Biosystéma,* no. 1: 1–126, Société française de systématique.

Michard, Jean-Guy. 1989. *Le Monde perdu des dinosaures.* Paris: Gallimard, collections "Découverte."

Norman, David. 1984. "A Systematic Reappraisal of the Reptile Order Ornithischia," in W. Reif and F. Westphal (eds.), *Third Symposium on Mesozoic Terrestrial Ecosystems, short papers.* Tübingen: Attempto Verlag, pp. 157–62.

    1986. *Illustrated Encyclopaedia of Dinosaurs.* London: Salamander.

Piveteau, Jean. 1968. "Cuvier 1769–1832," in *Encyclopaedia universalis,* vol. 5, pp. 254–6.

Sereno, Paul. 1984. "The Phylogeny of the Ornithischia: A Reappraisal," in W. Reif and F. Westphal (eds.), *Third Symposium on Mesozoic Terrestrial Ecosystems, short papers.* Tübingen: Attempto Verlag, pp. 219–26.

    1986. "Phylogeny of the Bird-Hipped Dinosaurs (Order Ornithischia)," *National Geographic Research,* 2: 234–56.

    1990. "Clades and Grades in Dinosaur Systematics," in Kenneth Carpenter and Philip J. Currie (eds.), *Dinosaur Systematics: Approaches and Perspectives.* Cambridge: Cambridge University Press, pp. 9–20.

Taquet, Philippe. 1975. "Remarques sur l'évolution des Iguanodontidés et l'origine des Hadrosauridés," *Problèmes actuels de paléontologie – évolution des vertébrés* (colloque international du CNRS), no. 218: 503–10.

    1976. *Géologie et paléontologie du gisement de Gadoufaoua (Aptien du Niger).* Paris: Éd. du CNRS, Cahiers de paléontologie.

    1977. "Dinosaurs of Niger," *Nigerian Field,* 42(1): 1–10.

    1994. "Georges Cuvier, ses liens scientifiques européens," in E. Buffetaut, J. M. Mazin, and E. Salmon (eds.), *Actes du Colloque international de Montbeliard.* Montbéliard: Société d'émulation de Montbéliard.

Tassy, Pascal. 1991. *L'Arbre à remonter le temps: Les Rencontres de la systématique et de l'évolution.* Paris: Christian Bourgois.

Weishampel, David. 1984. "Evolution of Jaw Mechanisms in Ornithopod Dinosaurs," *Advances in Anatomy, Embryology and Cell Biology,* 87: 1–110.

Weishampel, David, Peter Dodson, and Halszka Osmólska (eds.). 1990. *The Dinosauria.* Berkeley: University of California Press.

Weishampel, David, David Norman, and Dan Grigorescu. 1993. "*Telmatosaurus transsylvanicus* from the Late Cretaceous of Romania: The Most Basal Hadrosaurid Dinosaur," *Palaeontology,* 36(2): 361–85.

3. TRACKING THE DINOSAURS

Alexander, R. McNeill. 1989. *Dynamics of Dinosaurs and Other Extinct Giants.* New York: Columbia University Press.

———. 1993. "La Course des dinosaures," in *Les Dinosaures.* Paris: Pour la Science, dossier hors série, pp. 22–8.

Batory, Dana, and William A. S. Sarjeant. 1989. "Sussex *Iguanodon* and the Writing of *The Lost World*," in David D. Gillette and Martin G. Lockley (eds.), *Dinosaur Tracks and Traces.* Cambridge: Cambridge University Press, pp. 13–18.

Defretin, Suzanne, François Joulia, and Albert-Félix de Lapparent. 1956. "Les *Estheria* de la région d'Agadès (Niger)," *Bulletin de la Société géologique de France,* 6th ser., no. 6:. 679–90.

Gillette, David D., and Martin G. Lockley (eds.). 1989. *Dinosaur Tracks and Traces.* Cambridge: Cambridge University Press.

Ginsburg, Léonard, Albert-Félix de Lapparent, and Philippe Taquet. 1968. "Piste de *Chirotherium* dans le Trias du Niger," *Comptes rendus de l'Académie des sciences,* 266: 2056–8.

Ginsburg, Léonard, Albert-Félix de Lapparent, Bernard Loiret, and Philippe Taquet. 1966. "Empreintes de pas de vertébrés tétrapodes dans les séries continentales à l'ouest d'Agadès (République du Niger)," *Comptes rendus de l'Académie des sciences,* 263: 28–31.

Guérin-Franiatte, Simone, and Philippe Taquet. 1993. "Une Nouvelle Faune d'Estheries *(Branchiopoda, Conchostraca)* dans le Crétacé inférieur du nord-est du Brésil," *Documents du laboratoire de géologie de Lyon,* no. 125: 203–21.

Krebs, Bernard. 1965. "Die Triasfauna der Tessiner Kalkalpen, XIX, *Ticinosuchus ferox* nov. gen., nov. sp.," *Schweizerische Paläontologische Abhandlungen,* 81: 1–140.

Lapparent, Albert-Félix de, and Christian Montenat. 1967. "Les Empreintes de pas de reptiles de l'Infralias du Veillon (Vendée)," *Mémoires de la Société géologique de France,* new ser., 46(107): 1–44.

Ligabue, Giancarlo (ed.). 1984. *Sulle Orme dei Dinosauri,* Venice, Erizzo Editrice.

Lockley, Martin. 1991. *Tracking Dinosaurs, A New Look at an Ancient World.* Cambridge: Cambridge University Press.

Ostrom, John. 1972. "Were Some Dinosaurs Gregarious?" *Palaeogeography, Palaeoclimatology, Palaeoecology,* 11: 287–301.

Telles Antunes, Miguel. 1976. "Dinosáurios Eocretácicos de Lagosteiros," *Ciencias da Terra (Lisbon)*, 1: 1–35.

Thulborn, Tony. 1990. *Dinosaur Tracks*. London: Chapman & Hall.

4. MANY CROCODILES, ONE CONTINENT

Broin, France de. 1980. "Les Tortues de Gadoufaoua (Aptien du Niger). Aperçu sur la paléobiogéographie des *Pelomedusidae (Pleurodira)*," in *Écosystèmes continentaux du Mésozoïque, Mémoires de la Société géologique de France*, new ser., no. 139: 39–46.

Broin, France de, and Philippe Taquet. 1966. "Un Crocodilien nouveau dans le Crétacé inférieur du Sahara," *Comptes rendus de l'Académie des sciences*, 262: 2326–9.

Buffetaut, Éric, and Philippe Taquet. 1977. "The Giant Crocodilian *Sarcosuchus* in the Early Cretaceous of Brazil and Niger," *Palaeontology*, 20(1): 203–8.

1977. "Un crocodile géant à cheval sur deux continents," *La Recherche*, no. 76: 289–92.

1979. "An Early Cretaceous Terrestrial Crocodilian and the Opening of the South Atlantic," *Nature*, 280(5722): 486–7.

Colbert, Edwin H., and Roland T. Bird. 1954. "A Gigantic Crocodile from the Upper Cretaceous Beds of Texas," *American Museum Novitates*, no. 1688: 1–22.

Hallam, Anthony. 1979. "Alfred Wegener et l'hypothèse de la dérive des continents," in *La Dérive des continents, la tectonique des plaques*. Paris/Berlin: Bibliothèque Pour la Science, pp. 10–31.

Kipling, Rudyard. 1894. *The Jungle Book*. London / New York: MacMillan; repr. New York: Oxford University Press, World's Classics Series, 1987.

Lapparent, Albert-Félix de. 1947. "Stratigraphie et âge du continental intercalaire dans le Gourara, le Touat et le Tidikelt (Sahara algérien)," *Comptes rendus de l'Académie des sciences*, 225: 754–6.

Mawson, Joseph, and Arthur Smith Woodward. 1907. "On the Cretaceous Formation of Bahia (Brazil), and on Vertebrate Fossils Collected Therein," *Quarterly Journal of the Geological Society*, 62: 128–38.

Wegener, Alfred. 1928. *Die Entstehung der Kontinente und Ozeane;* trans. as *La Genèse des continents et des océans*. Paris: Nizet et Bastard, 1937; repr. Paris: Christian Bourgois, 1990.

Wenz, Sylvie. 1975. "Un Nouveau Cœlacanthide du Crétacé inférieur du Niger: Remarques sur l'évolution des Iguanodontides et l'origine des Hadrosaurides," *Problèmes actuels de Paléontologie – Evolution des vertébrés* (colloque international du CNRS), no. 218: 503–10.

5. IN MOROCCO WITH THE GIANTS OF THE ATLAS

Bakker, Robert T. 1972. "Anatomical and Ecological Evidence of Endothermy in Dinosaurs," *Nature*, 238: 81–5.

1990. *The Dinoscaur Heresies*. New York: William Morrow & Co.

Bourcart, Jacques, Albert-Félix de Lapparent, and Henri Termier. 1942. "Un Nouveau Gisement de Dinosauriens jurassiques au Maroc," *Comptes rendus de l'Académie des sciences*, 214: 120–2.

Charrière, André. 1992. "Discontinuités entre les 'couches rouges' du Jurassique moyen et du Crétacé inférieur dans le Moyen-Atlas (Maroc)," *Comptes rendus de l'Académie des sciences*, 315: 1389–96.

Colbert, Edwin H. 1968. *Men and Dinosaurs*. New York: Dutton.

Dercourt, Jean, Luc Emmanuel Ricou, and Bruno Vrielynck. 1993. *Atlas Téthys Palaeoenvironmental Map*. Paris: Gauthier–Villars.

Dercourt, Jean, and Jacques Paquet. 1985. *Géologie: Objets et méthodes*. Paris: Dunod.

Dresnay, Renaud du. 1963. "Données stratigraphiques complémentaires sur le Jurassique moyen des synclinaux d'El Mers et de Skoura (Moyen-Atlas, Maroc)," *Bulletin de la Société géologique de France*, 7th ser., no. 5: 883–900.

Dutuit, Jean-Michel, and Ahmed Ouazzou. 1980. "Découverte d'une piste de dinosaure sauropode sur le site d'empreintes de Demnat (Haut Atlas marocain)," *Écosystèmes continentaux du Mésozoïque, Mémoires de la Société géologique de France*, new ser., no. 139: 95–102.

Ellenberger, François. 1987. "Un Centenaire à commémorer: La Découverte des charriages de Provence par Marcel Bertrand," *Travaux du Comité français d'histoire de la géologie (Cofrhigéo)*, 3d ser., 1(6): 49–56.

Greppin, Jean-Baptiste. 1870. "Description géologique du Jura bernois et de quelques districts adjacents," *Beitrage Geologische Karte Schweiz*, no. 8: 1–339.

Huene, Friedrich von. 1922. "Über einen Sauropoden im oberen Malm des Berner Jura," *Eclogae Geologiae Helvetiae*, 17: 80–94.

Jenny, Jacques, Catherine Jenny-Deshusses, Alain Le Marrec, and Philippe Taquet. 1980. "Découverte d'ossements de dinosauriens dans le Jurassique inférieur (Toarcien) du Haut-Atlas central (Maroc)," *Comptes rendus de l'Académie des sciences*, 290: 839–42.

Lapparent, Albert-Félix de. 1955. *Étude paléontologique des vertébrés du Jurassique d'El Mers (Moyen-Atlas)*, notes et memoires, no. 124. Rabat: Éd. du Service géologique du Maroc.

Lévi-Strauss, Claude. 1973. *Tristes tropiques*. Paris: Plon, collection "Terre humaine."

1989. *Des Symboles et leurs doubles*. Paris: Plon.

MacIntosh, J. S. 1990. "Species Determination in Sauropod Dinosaurs with Tentative Suggestions for Their Classification," in Kenneth Carpenter and Philip J. Currie (eds.), *Dinosaur Systematics: Approaches and Perspectives*. Cambridge: Cambridge University Press, pp. 53–69.

Monbaron, Michel. 1983. "Dinosauriens du Haut-Atlas central (Maroc): État de recherche et précision sur la découverte d'un squelette complet de grand Cétiosaure," *Actes de la Société jurassienne d'émulation*, 203–34.

Monbaron, Michel, and Philippe Taquet. 1981. "Découverte du squelette complet d'un grand Cétiosaure (dinosaure sauropode) dans le bassin jurassique moyen de Tilougguit (Haut-Atlas central, Maroc)," *Comptes rendus de l'Académie des sciences*, 292: 243–6.

Ostrom, John H. 1969. "Osteology of *Deinonychus antirrhopus*, an Unusual Theropod from the Lower Cretaceous of Montana," *Peabody Museum of Natural History Bulletin*, 30: 1–165.

1970. "Stratigraphy and Paleontology of the Cloverly Formation (Lower Cretaceous) of the Bighorn Basin Area, Wyoming and Montana," *Peabody Museum of Natural History Bulletin*, 35: 1–234.

Plateau, Henri, Germain Giboulet, and Édouard Roch. 1937. "Sur la présence d'empreintes de Dinosauriens dans la région de Demnat (Maroc)," *Comptes rendus sommaires de la Société géologique de France*, 1937: 241–2.

6. IN THE STEPPES OF CENTRAL ASIA

Andrews, Roy Chapman. 1932. "The New Conquest of Central Asia: A Narrative of the Exploration of the Central Asiatic Expeditions in Mongolia and China," in *Natural History of Central Asia*, vol. 1, New York: American Museum of Natural History.

Bacot, Jacques. 1912. *Le Tibet révolté*. Paris: Hachette; repr. Paris: Raymond Chabaud, 1988.

Bailly, André, Eric Buffetaut, Gilles Cheylan, Marc Godinot, and Monique Vianey-Liaud (eds.). 1994. *Dinosaures en Provence*. Aix-en-Provence: Muséum d'Histoire naturelle d'Aix-en-Provence, Cahiers de Sainte Victoire.

Barsbold, Rinchen. 1992. "Les Dinosaures de Mongolie," in *Dinosaures et mammifères du désert de Gobi* (exhib. vol.). Paris: Éd. du Muséum Nationale d'Histoire naturelle, pp. 63–75.

Efremov, I. A. 1954. *Récits – contes scientifiques*. Moscow: Foreign Language Editions.

Fournier, R., and Philippe Matheron. 1983. *Soixante ans de géologie provençale*, Musée d'Histoire naturelle de Marseille.

Horner, John. 1982. "Evidence for Colonial Nesting and 'Site Fidelity' among Ornithischian Dinosaurs," *Nature*, 297: 675–6.

1993. "Les Œufs et les nids de dinosaures," in *Les Dinosaures*. Paris: Pour la Science, dossier hors série, pp. 14–21.

Kielan-Jaworowska, Zofia, and Rinchen Barsbold. 1972. "Narrative of the Polish–Mongolian Palaeontological Expeditions 1967–1971," *Palaeontologia Polonica*, 27: 5–13.

Ligabue, Giancarlo. 1992. "Des dragons aux dinosaures," in *Dinosaures et mammifères du désert de Gobi* (exhib. vol.). Paris: Éd. du Muséum Nationale d'Histoire naturelle, pp. 31–4.

Prejwalski, Nicolas de. "Voyage en Mongolie et au pays des Tangoutes," in *Le Tour du monde*. Paris: Hachette, 1877.

Rojdestvenski, Anatoli. 1960. "Chasse aux dinosaures dans le desert de Gobi," Paris: Fayard.

Seymour, Roger S. 1980. "Dinosaur Eggs: The Relationships between Gas Conductance through the Shells, Water Loss during Incubation, and

Clutch Size," in *Écosystèmes continentaux du Mésozoïque, Mémoires de la Société géologique de France*, new ser., no. 139: 177–84.

Sigogneau-Russell, Denise. 1991. *Les Mammifères au temps des dinosaures.* Paris: Masson, collection "Les Grands problèmes de l'evolution."

———. 1992. "La Longue Marche des mammifères mésozoïques," in *Dinosaures et mammifères du désert de Gobi* (exhib. vol.). Paris: Éd. du Muséum Nationale d'Histoire naturelle, pp. 77–91.

Taquet, Philippe. 1992. "À la recherche des dinosaures du desert de Gobi, 1921–1991," in *Dinosaures et mammifères du désert de Gobi* (exhib. vol.). Paris: Éd. du Muséum Nationale d'Histoire naturelle, pp. 15–19.

Zorzi, Alvis, Viviano Domenici, Antonio Paolillo, Ivana Grollova, Adriano Madaro, David Sneath, N. Tsultem, Giancarlo Ligabue, Philippe Taquet, and Rinchen Barsbold. 1992. *Mongolia, Nelle Steppe di Cinghis Khaan.* Venice: Erizzo Editrice.

7. A BONE HUNTER IN LAOS

Bois, G., and Josué-Heilman Hoffet. 1949. "Entraîneur d'hommes," *Éducation*, Haut Commissariat de France pour l'Indochine, 2(14): 1–6.

Fromaget, Jacques. 1942. "La Question des grès supérieurs en Indochine," *Comptes rendus des séances du Conseil des recherches scientifiques de l'Indochine*, Hanoi, 19(1): 168.

Hoffet, Josué-Heilmann. 1933. "Étude géologique sur le centre de l'Indochine entre Tourane et le Mekong (Annam central et Bas Laos)," *Bulletin du Service géologique de l'Indochine*, Hanoi, 20(fasc. 2).

———. 1933. "Les Mois de la chaîne annamitique entre Tourane et les Boloven, terre-air-mer," *La Géographie*, Société de géographie, Paris, 59(1): 1–43.

———. 1936. "Découverte du Crétacé en Indochine," *Comptes rendus de l'Académie des sciences*, 202: 1–2.

———. 1937. "Note sur la géologie du Bas-Laos," *Bulletin du Service géologique de l'Indochine*, Hanoi (supplement to 20[fasc. 2]), 24(fasc. 2).

———. 1943. "Description de quelques ossements de titanosauriens du Sénonien du Bas-Laos," *Comptes rendus des séances du Conseil des recherches scientifiques de l'Indochine*, Hanoi, 20(1): 49–57.

———. 1944a. "Description des ossements les plus caractéristiques appartenant à des Avipelviens du Sénonien du Bas-Laos," *Comptes rendus des séances du Conseil des recherches scientifiques de l'Indochine*, Hanoi, 21(1), 8 pp. (repr.).

———. 1944b. "L'Age des formations a dinosaures du Bas-Laos," *Comptes rendus des séances du Conseil des recherches scientifiques de l'Indochine*, Hanoi, 21(1), 5 pp. (repr.).

Reyberol, Yvonne. 1993. "Les Dinosaures du Laos, un vrai feuilleton," *Le Monde* (Paris), January 13.

Riabinin, A. N. 1930. "*Mandschurosaurus Amurensis* nov. gen. nov. sp., a Hadrosaurian Dinosaur from the Upper Cretaceous of Amur River," *Monogr. Russ. Paleontol. Obschest.*, 2: 1–33.

Taquet, Philippe. 1993. "À la recherche des dinosaures du Laos," *Les Amis du Muséum,* no. 173: 1–2.

Taquet, Philippe, Bernard Battail, and Jean Dejax. 1992. "New Discoveries of Sauropod and Ornithopod Dinosaurs in the Lower Cretaceous of Laos," *Journal of Vertebrate Palaeontology,* 12(3): 55A (abst.).

Vidal, Jules. 1972. *La Végétation du Laos, I. Le Milieu (conditions écologiques).* Toulouse: Éd. Vithagna, collection "Documents pour le Laos."

8. ACROSS EUROPE WITH THE DINOSAURS

Bigot, A. 1944. "À propos d'un portrait de William Buckland," *Bulletin de la Société linnéenne de Normandie,* 9th ser., 3: 130–1.

Buckland, William. 1824. "Notice of the *Megalosaurus* or Great Fossil Lizard of Stonesfield," *Transactions of the Geological Society,* London, 2d ser., 1(21): 390–6.

Charig, Alan J., and Angela Milner. 1986. "*Baryonyx,* a Remarkable New Theropod Dinosaur," *Nature,* 324: 359–61.

Cuvier, Georges. 1798–9. "Mémoires sur les espèces d'éléphants vivantes et fossiles," *Mémoires de l'Institut,* Paris, 2: 1–22; repr. *Journal de physique,* Paris, 50: 207–17, 1800.

——— 1808. "Sur le grand animal fossile des carrières de Maestricht," *Annales du Muséum d'histoire naturelle,* Paris, 12: 145–76.

——— 1808. "Sur les ossemens fossiles de crocodiles et particulièrement ceux de la region du Havre et de Honfleur, avec des remarques sur les squelettes des sauriens de la Thuringe, article IV: Description des ossemens des environs de Honfleur et du Havre; leur comparaison avec ceux du gavial; détermination des deux espèces inconnues de crocodiles qui les ont fournies," *Annales du Muséum d'histoire naturelle,* Paris, 73–110.

——— 1812. *Discours préliminaire à l'ouvrage: Recherches sur les ossemens fossiles de quadrupèdes ou l'on rétablit les caractères de plusieurs espèces d'animaux que les révolutions du globe paraissent avoir détruites.* Paris: Deterville, 4 vols., 1821–4; repr. Paris: Christian Bourgois, 1985, with a preface by Herbert Thomas and afterword by Goulven Laurent.

——— 1824. *Recherches sur les ossemens fossiles,* 2d ed., vol. 5, pt. 2. Paris: Dufour et d'Ocagne.

Desmond, Adrian. 1975. *The Hot-Blooded Dinosaurs.* London: Blond & Briggs.

Dicquemare. 1776. "Observations sur la physique, sur l'histoire naturelle et sur les arts, avec des planches en taille-douce dédiées à Monsieur le comte d'Artois," *Journal de physique,* Paris, 7: 404–14.

Dughi, Raymond, and François Sirugue. 1957. "Les Œufs de dinosaures du bassin d'Aix-en-Provence," *Comptes rendus de l'Académie des sciences,* 245: 707–10.

——— 1960. "Les Dinosaures vivaient en Provence au Maastrichtien (bégudien)," *Comptes rendus de l'Académie des sciences,* 251: 2387–9.

Eudes-Deslongchamps, Jacques Amand. 1837. *Mémoire sur le* Poekilopleuron bucklandii, *grand saurien fossile intermédiaire entre les crocodiles et les lézards découvert dans les carrières de la Maladrerie près de Caen, au mois de juillet 1837.* Caen: Hardel. (This text was published in 1838 in the *Memoirs of the Linnean Society of Normandy,* 6: 37–146.)

Hecht, Max K., John H. Ostrom, Gunter Viohl, and Peter Wellnhofer (eds.). 1985. *The Beginnings of Birds: Proceedings of the International Archaeopteryx Conference Eichstätt, 1984.* Eichstätt: Freunde des Jura-Museums.

Mantell, Gideon. 1822. *The Fossils of the South Downs; or, Illustrations of the Geology of Sussex.* London: L. Relfe.

1825. "Notice on the *Iguanodon,* a Newly Discovered Fossil Reptile from the Sandstone of Tilgate in Sussex," *Philosophical Transactions of the Royal Society of London,* 111(1): 179–86.

1833. *The Geology of the South East of England.* London: Longman.

Mathiot, Charles, and M. Duvernoy. 1940. "Lettres inédites de Charles Laurillard à Georges-Louis Duvernoy, Introduction et notes par le pasteur Charles Mathiot et le docteur M. Duvernoy," *Mémoires de la Société d'émulation de Montbéliard,* 55: 3–48.

Meyer, Hermann von. 1832. *Palaeologica zur Geschichte der Erde und ihrer Geschöpfe.* Frankfurt-am-Main: S. Schmerber.

Ostrom, John. 1976. "*Archaeopteryx* and the Origin of Birds," *Biological Journal of the Linnean Society,* 8(2): 91–182.

Owen, Richard. 1842. "Report on British Fossil Reptiles," pt. 2, *Reports of the British Association for the Advancement of Science,* Plymouth, 1842: 60–204.

Pfaff, C. M. 1858. *Lettres de Georges Cuvier à C. M. Pfaff, sur l'histoire naturelle, la politique et la littérature – 1788–1792,* trans. Louis Marchant. Paris: Masson.

Sanz, José-Luis, Jose F. Bonaparte, and A. Lacasa Ruiz. 1988. "Unusual Early Cretaceous Birds from Spain," *Nature,* 331(6155): 433–5.

Stromer, Ernst. 1915. "Wirbeltier-Reste der Baharije-Stufe (unterstes Cenoman), 3. Das Original des Theropoden *Spinosaurus aegyptiacus* nov. gen., nov. sp.," *Abhandlungen der Königliche Bayerische Akademie der Wissenschaften, Math.-Phys.,* 28: 1–32.

Taquet, Philippe. 1983. "Cuvier–Buckland–Mantell et les dinosaures," *Actes du symposium paléontologique G. Cuvier.* Montbéliard: Société d'émulation de Montbéliard, pp. 475–94.

1984. "Une curieuse spécialisation du crâne de certains dinosaures carnivores du Crétacé: Le Museau long et étroit des Spinosaurides," *Comptes rendus de l'Académie des sciences,* ser. 2, 299(5): 217–22.

1994. "Georges Cuvier, ses liens scientifiques européens," in E. Buffetaut, J. M. Mazin, and E. Salmon (eds.), *Actes du Colloque international de Montbeliard.* Montbéliard: Société d'émulation de Montbéliard.

Wellnhofer, Peter. 1993. "L'Archéoptéryx," in *Les Dinosaures.* Paris: Pour la Science, dossier hors série, pp. 60–73.

9. THE RISE AND FALL OF THE DINOSAURIAN EMPIRE

Alvarez, Luis W., Walter Alvarez, Frank Asaro, and Helen Michel. 1980. "Extraterrestrial Cause for the Cretaceous–Tertiary Extinction," *Science,* 208: 1095–108.

Archibald, David. 1987. "A Reassessment of Reptilian Diversity across the Cretaceous–Tertiary Boundary," *Natural History Museum of Los Angeles County Contributions in Science,* no. 391: 1–26.

Benton, Michael. 1990. "Origin and Interrelationships of Dinosaurs," in David Weishampel, Peter Dodson, and Halszka Osmólska (eds.), *The Dinosauria,* Berkeley: University of California Press, pp. 11–30.

Bonis, Louis de. 1991. *Évolution et extinction dans le regne animal.* Paris: Masson.

Clemens, William A. 1986. "Evolution of the Terrestrial Vertebrate Fauna during the Cretaceous–Tertiary Transition," in D. K. Elliott (ed.), *The Dynamics of Extinction,* New York: Wiley–Interscience, pp. 63–86.

Courtillot, Vincent, J. Besse, D. Vandamme, J.-J. Jaeger, and R. Montigny. 1986. "Les Épanchements volcaniques du Deccan (Inde), cause des extintions biologiques à la limite Crétacé–Tertiaire?" *Comptes rendus de l'Académie des sciences,* ser. 2, 303: 863–8.

Cuvier, Georges. *Discours sur les révolutions de la surface du globe,* 1825. Paris: Dufour.

Darwin, Charles. 1859. *On the Origin of Species.* London: J. Murray.

Dodson, Peter. 1990. "Counting Dinosaurs: How Many Kinds Were There?" *Proceedings of the National Academy Sciences of the USA,* 87: 7608–12.

Dodson, Peter, and Susan Dawson. 1991. "Making the Fossil Record of Dinosaurs," *Modern Geology,* 16 (1–2): 3–15.

Frakes, Larry A. 1978. *Climates Throughout Geologic Time.* Amsterdam: Elsevier.

Frakes, Larry A., Jean-Luc Probst, and Wolfgang Ludwig. 1994. "Latitudinal Distribution of Paleotemperature on Land and Sea from Early Cretaceous to Middle Miocene," *Comptes rendus de l'Académie des sciences,* ser. 2, 318: 1209–18.

Gauthier, Jacques. 1986. "Saurischian Monophyly and the Origin of Birds," in Kevin Padian (ed.), *The Origin of Birds and the Evolution of Flight.* San Francisco: Memoirs of the California Academy of Sciences, no. 8. pp. 1–55.

Gibbon, Edward. 1776–88. *The History of the Decline and Fall of the Roman Empire.* Repr. New York: Modern Library, 1932.

Ginsburg, Léonard. 1964. "Les Régressions marines et le problème du renouvellement des faunes au cours des temps géologiques," *Bulletin de la Société géologique de France,* 7th ser., 6(1): 13–20.

    1984. "Théories scientifiques et extinction des dinosaures," *Comptes rendus de l'Académie des sciences,* ser. 2, 298(7): 317–20.

Hecht, Max M., and Antoni Hoffmann. 1986. "Why Not Neo-Darwinism? A Critique of Paleobiological Challenges," *Oxford Surveys in Evolutionary Biology,* 3: 1–4.

Hooykaas, Reijer. 1970. *Continuité et discontinuité en géologie et biologie*. Paris: Éd. du Seuil.

Jablonski, David. 1986. "Causes and Consequences of Mass Extinctions: A Comparative Approach," in D. K. Elliott (ed.), *Dynamics of Extinction*. New York: Wiley–Interscience, pp. 183–230.

Jacob, François. 1970. *La Logique du vivant*. Paris: Gallimard.

Keller, Gerta, E. Barbara, B. Schmitz, and E. Mattson. 1993. "Gradual Mass Extinction, Species Survivorship and Long-Term Environmental Changes across the Cretaceous–Tertiary Boundary in High Latitudes," *Geological Society of America Bulletin*, 105: 979–99.

Laurent, Goulven. 1987. *Paléontologie et évolution en France, 1800–1860: De Cuvier–Lamarck à Darwin*. Paris: Éd. du Comité des travaux historiques et scientifiques.

Lyell, Charles. 1830–3. *Principles of Geology; Being an Attempt to Explain the Former Changes of the Earth's Surface by Reference to Causes Now in Operation*, vols. 1–3. London: J. Murray.

Officer, Charles B., and Charles L. Drake. 1983. "The Cretaceous–Tertiary Transition," *Science*, 219: 1383–90.

Piveteau, Jean. 1962. "L'Origine de l'homme et la paléontologie," in *L'Evolution, les conférences de la Maison des sciences*, Association cooperative des étudiants en sciences de Paris. Paris: Masson, pp. 421–36.

Rocchia, Robert, Daniel Boclet, Philippe Bonte, Laurence Froget, Bruno Galbrun, Celestine Jehanno, and Eric Robin. 1992. "Iridium and Other Element Distributions, Mineralogy and Magnetostratigraphy Near the Cretaceous/Tertiary Boundary Inhole 761c. Proceedings of the Ocean Drilling Program," *Scientific Results*, 122: 753–62.

Stanley, Steven M. 1987. *Extinctions*. New York: Scientific American Library.

Taquet, Philippe. 1993. "Les Dinosaures, grandeur et decadence," *La Vie des sciences, comptes rendus*, série générale, 10(4): 265–84.

Tassy, Pascal. 1991. *Le Message des fossiles*. Paris: Hachette.

Van Valen, Leigh, and Robert E. Sloan. 1977. "Ecology and the Extinction of the Dinosaurs," *Evolutionary Theory*, 2: 37–64.

Wallace, Malcolm W., Reid R. Keays, and Victor A. Gostin. 1991. "Stromatolitic Iron Oxides: Evidence That Sea-Level Changes Can Cause Sedimentary Iridium Anomalies," *Geology*, 19(6): 551–4.

INDEX

For ECONOMY, informal and formal names of taxa (titanosaur, Titano-sauridae) as well as eponymous genera (*Titanosaurus*) are often listed under a single joint entry. Illustration pages are followed by an "f."

■

238